A SHORT INTRODUCTION TO QUANTUM INFORMATION AND QUANTUM COMPUTATION

Quantum information and computation is a rapidly expanding and cross-disciplinary subject. This book gives a self-contained introduction to the field for physicists, mathematicians and computer scientists who want to know more about this exciting subject. After a step-by-step introduction to the quantum bit (qubit) and its main properties, the author presents the necessary background in quantum mechanics. The core of the subject, quantum computation, is illustrated by a detailed treatment of three quantum algorithms: Deutsch, Grover and Shor. The final chapters are devoted to the physical implementation of quantum computers, including the most recent aspects, such as superconducting qubits and quantum dots, and to a short account of quantum information.

Written at a level suitable for undergraduates in physical sciences, no previous knowledge of quantum mechanics is assumed, and only elementary notions of physics are required. The book includes many short exercises, with solutions available to instructors through solutions@cambridge.org.

MICHEL LE BELLAC is Emeritus Professor at the University of Nice, and a well known elementary particle theorist. He graduated from Ecole Normale Supérieure in 1962, before conducting research with CNRS. In 1967 he returned to the University of Nice, and was appointed Full Professor of Physics in 1971, a position he held for over 30 years. His main fields of research have been the theory of elementary particles and field theory at finite temperatures. He has published four other books in French and three other books in English, including *Thermal Field Theory* (Cambridge, 1996), *Equilibrium and Non-Equilibrium Statistical Thermodynamics* with Fabrice Mortessagne and G. George Batrouni (Cambridge, 2004) and *Quantum Physics* (Cambridge, 2006).

A SHORT INTRODUCTION
TO QUANTUM INFORMATION
AND QUANTUM COMPUTATION

Michel Le Bellac

Translated by Patricia de Forcrand-Millard

CAMBRIDGE
UNIVERSITY PRESS

CAMBRIDGE UNIVERSITY PRESS
Cambridge, New York, Melbourne, Madrid, Cape Town, Singapore, São Paulo

Cambridge University Press
The Edinburgh Building, Cambridge CB2 2RU, UK

Published in the United States of America by Cambridge University Press, New York

www.cambridge.org
Information on this title: www.cambridge.org/9780521860567

French edition © Editions Belin 2005
English translation © Cambridge University Press 2006

First published as *Introduction à l'information quantique* by Editions Belin 2005
First published in English 2006

Printed in the United Kingdom at the University Press, Cambridge

A catalog record for this publication is available from the British Library

Library of Congress Cataloging in Publication data

ISBN-13 978-0-521-86056-7 hardback
ISBN-10 0-521-86056-3 hardback

Contents

Foreword

Quantum physics is well known for being counter-intuitive, or even bizarre. Quantum correlations have no equivalence in classical physics. This was all well known for years. But several discoveries in the 1990's changed the world. First, in 1991 Artur Ekert, from Oxford University, discovered that quantum correlations could be used to distribute cryptographic keys. Suddenly physicists realized that quantum correlations and its associated bizarre non-locality, could be exploited to achieve a useful task that would be impossible without quantum physics. What a revolution! And this was not the end. Three years later, Peter Shor, from the AT&T Laboratories, discovered an algorithm that breaks the most used public key crypto-systems. Shor's algorithm requires a quantum computer, yet another bizarre quantum device, a kind of computer that heavily exploits the quantum superposition principle. The following year, in 1995, a collaboration between six physicists and computer scientists from three continents led to the discovery of quantum teleportation, a process with a science-fiction flavour.

These discoveries and others led to the emergence of a new science, marrying quantum physics and theoretical computer science, called quantum information science. Today, a steadily growing community of physicists, mathematicians and computer scientists develop the tools of this new science. This leads to new experiments and new insights into physics and information theory. It is a fascinating time.

Quantum information is still a very young science. In particular, the technology required to build a quantum computer is still unknown. Only quantum key distribution has reached a certain level of maturity, with a few start-ups already offering complete systems. Nevertheless, quantum information has been widely recognised as a source for truly innovative ideas and disruptive technologies.

In this context Michel Le Bellac's book is especially welcome. It provides a concise, yet precise, introduction to quantum information science. The book will be of great help to anyone who would like to understand the basics of quantum information. It is accessible to non-physicists and to physicists not at ease with quantum oddities. It will contribute to bringing together the different communities that are joining forces to develop this science of the future. For this purpose one needs to get used to concepts such as entanglement, non-locality, teleportation and superpositions of seemingly exclusive state of affairs, etc. A demanding but very joyful exploration for all readers!

Nicolas Gisin
Geneva, November 2005

Preface

This book originated in a course given first to computer scientists at the University of Nice–Sophia Antipolis in October 2003, and later to second-year graduate students at the same University.

Quantum information is a field which at present is undergoing intensive development and, owing to the novelty of the concepts involved, it seems to me it should be of interest to a broad range of scientists beyond those actually working in the field. My goal here is to give an elementary introduction which is accessible not only to physicists, but also to mathematicians and computer scientists desiring an initiation into the subject. This initiation includes the essential ideas from quantum mechanics, and the only prerequisite is knowledge of linear algebra at the undergraduate level and some physics background at the high school level (except in Chapter 6 where I discuss the physical implementations of quantum computers and more physics background is required). In order to make it easier for mathematicians and computer scientists to follow, I give an elementary presentation of the Dirac notation. Quantum physics is an extremely large subject, and so I have attempted to limit the introductory concepts to the barest minimum needed to understand quantum information.

In Chapter 2, I introduce the concept of the quantum bit or qubit using the simplest possible example, that of the photon polarization. This example allows me to present the essential ideas of quantum mechanics and explain quantum cryptography. Chapter 3 generalizes the concept of qubit to other physical systems like spin 1/2 and the two-level atom, and explains how qubits can be manipulated by means of Rabi oscillations. Quantum correlations, which are introduced in Chapter 4, certainly represent the situation where the classical and quantum concepts most obviously diverge. Several ideas such as entanglement and the concept of state operator (or density operator) which are essential for what follows are introduced in the same chapter.

After dealing with the indispensable preliminaries, in Chapter 5 I reach the core of the subject, quantum computing. The quantum logic gates are used to construct quantum logic circuits which allow specific algorithms to be realized and illustrate quantum parallelism. Three algorithms are explained in detail: the Deutsch algorithm, the Grover search algorithm, and the Shor factorization algorithm. Chapter 6 describes four possible physical realizations of a quantum computer: computers based on NMR, trapped ions, Josephson junctions (superconducting qubits), and quantum dots. Finally, in Chapter 7 I give a brief introduction to quantum information: teleportation, the von Neumann entropy, and quantum error-correction codes.

I am very grateful to Joël Leroux of the Ecole Supérieure in Computing Sciences of Sophia Antipolis for giving me the opportunity to teach this course. I also thank Yves Gabellini for rereading the manuscript and Jean-Paul Delahaye for his comments and especially for his essential aid in the writing of Section 5.9. This book has come into existence thanks to the support of Patrizia Castiglione and Annick Lesne of Éditions Belin, and also Simon Capelin and Vince Higgs of Cambridge University Press, whom I take this opportunity to thank. Finally, I am very grateful to Patricia de Forcrand-Millard for the excellence of her translation and to Nicolas Gisin for his foreword.

1
Introduction

Quantum information is concerned with using the special features of quantum physics for the processing and transmission of information. It should, however, be clearly understood that any physical object when analyzed at a deep enough level is a quantum object; as Rolf Landauer has succinctly stated, "A screwdriver is a quantum object." In fact, the conduction properties of the metal blade of a screwdriver are ultimately due to the quantum properties of electron propagation in a crystalline medium, while the handle is an electrical insulator because the electrons in it are trapped in a disordered medium. It is again quantum mechanics which permits explanation of the fact that the blade, an electrical conductor, is also a thermal conductor, while the handle, an electrical insulator, is also a thermal insulator. To take an example more directly related to information theory, the behavior of the transistors etched on the chip inside your computer could not have been imagined by Bardeen, Brattain, and Shockley in 1947 were it not for their knowledge of quantum physics. Although your computer is not a quantum computer, it does function according to the principles of quantum mechanics!

This quantum behavior is also a *collective* behavior. Let us give two examples. First, if the value 0 of a bit is represented physically in a computer by an uncharged capacitor while the value 1 is represented by the same capacitor charged, the passage between the charged and uncharged states amounts to the displacement of 10^4 to 10^5 electrons. Second, in a classic physics experiment, sodium vapor is excited by an electric arc, resulting in the emission of yellow light, the well known "yellow line of sodium." However, it is not actually the behavior of an individual atom that is observed, as the vapor cell typically contains 10^{20} atoms.

The great novelty since the early 1980s is that physicists now know how to *manipulate and observe individual quantum objects* – photons, atoms, ions, and so on – and not just the collective quantum behavior of a large number of such objects. It is this possibility of manipulating and observing individual quantum objects which lies at the foundation of quantum computing, as these quantum objects can be used as the physical support for quantum bits. Let us emphasize, however, that no new fundamental concept has been introduced since the 1930s. If the founding fathers of quantum mechanics (Heisenberg, Schrödinger, Dirac, . . .) were resurrected today, they would find nothing surprising in quantum information, but they would certainly be impressed by the skills of experimentalists, who have now learned how to perform experiments which in the past were referred to as "gedanken experiments" or "thought experiments."

It should also be noted that the ever-increasing miniaturization of electronics will eventually be limited by quantum effects, which will become important at scales of tens of nanometers. *Moore's law* [1] states that the number of transistors which can be etched on a chip doubles every 18 months, leading to a doubling of the memory size and the computational speed (amounting to a factor of 1000 every 15 years!). The extrapolation of Moore's law to the year 2010 implies that the characteristic dimensions of circuits on a chip will reach a scale of the order of 50 nanometers, and somewhere below 10 nanometers (to be reached by 2020?) the individual properties of atoms and electrons will become predominant, so that Moore's law may cease to be valid ten to fifteen years from now.

Let us take a very preliminary look at some characteristic features of quantum computing. A classical bit of information takes the value 0 or 1. A quantum bit, or *qubit*, can not only take the values 0 and 1, but also, in a sense which will be explained in the following chapter, all intermediate values. This is a consequence of a fundamental property of quantum states: it is possible to construct linear superpositions of a state in which the qubit has the value 0 and of a state in which it has the value 1.

The second property on which quantum computing is based is *entanglement*. At a quantum level it can happen that two objects form a single entity, even at arbitrarily large separation from each other. Any attempt to view this entity as a combination of two independent objects fails, unless the possibility of signal propagation at superluminal speeds is allowed. This conclusion follows from the theoretical work of John Bell in 1964, inspired by the studies of Einstein, Podolsky, and Rosen (EPR) in 1935, and from the experiments motivated by these studies (see Section 4.5 below). As we shall see in Chapter 5, the amount of information contained in an entangled state of N qubits grows exponentially with N, and not linearly as in the case of classical bits.

The combination of these two properties, linear superposition and entanglement, lies at the core of *quantum parallelism*, the possibility of performing a large number of operations in parallel. However, the principles of quantum parallelism differ fundamentally from those of classical parallelism. Whereas in a classical

---

[1] Moore's law is not a law based on theory, but rather an empirical statement which has been observed to hold over the last forty years.

computer it is always possible to know (at least in principle) what the internal state of the computer is, such knowledge is *in principle* impossible in a quantum computer. Quantum parallelism has led to the development of entirely new algorithms such as the Shor algorithm for factoring large numbers into primes, an algorithm which by its nature cannot be run on a classical computer. It is in fact this algorithm which has stimulated the development of quantum computing and has opened the door to a new science of algorithms.

Quantum computing opens up fascinating perspectives, but its present limitations should also be emphasized. These are of two types. First, even if quantum computers were available today, the number of algorithms of real interest is at present very limited. However, there is nothing which prevents others from being imagined in the future. The second type of limitation is that we do not know if it will someday be possible to construct quantum computers large enough to manipulate hundreds of qubits. At present, we do not know what the best physical support for qubits will be, and we know at best how to manipulate only a few qubits (a maximum of seven; see Chapter 6). The Enemy Number One of a quantum computer is *decoherence*, the interaction of qubits with the environment which blurs the delicate linear superpositions. Decoherence introduces errors, and ideally a quantum computer must be completely isolated from its environment. This in practice means the isolation must be good enough that any errors introduced can be corrected by error-correcting codes specific to qubits.

In spite of these limitations, quantum computing has become the passion of hundreds of researchers around the world. This is cutting-edge research, particularly that on the manipulation of individual quantum objects. This work, in combination with entanglement, permits us to speak of a "new quantum revolution" which is developing into a veritable quantum engineering. Another application might be the building of computers designed to simulate quantum systems. And, as has often happened in the past, such fundamental research may also result in new applications completely different from quantum computing, applications which we are not in a position to imagine today.

2

What is a qubit?

2.1 The polarization of light

Our first example of a qubit will be the polarization of a photon. First we briefly review the subject of light polarization. The *polarization of light* was demonstrated for the first time by the Chevalier Malus in 1809. He observed the light of the setting sun reflected by the glass of a window in the Luxembourg Palace in Paris through a crystal of Iceland spar. He showed that when the crystal was rotated, one of the two images of the sun disappeared. Iceland spar is a birefringent crystal which, as we shall see below, decomposes a light ray into two rays polarized in perpendicular directions, while the ray reflected from the glass is (partially) polarized. When the crystal is suitably oriented one then observes the disappearance (or strong attenuation) of one of the two rays. The phenomenon of polarization displays the vector nature of light waves, a property which is shared by shear sound waves: in an isotropic crystal, a sound wave can correspond either to a vibration transverse to the direction of propagation, i.e., a shear wave, or to a longitudinal vibration, i.e., a compression wave. In the case of light the vibration is only transverse: the electric field of a light wave is orthogonal to the propagation direction.

Let us recall the mathematical description of a planar and monochromatic scalar wave traveling in the z direction. The amplitude of vibration $u(z, t)$ as a function of time t has the form

$$u(z, t) = u_0 \cos(\omega t - kz),$$

where ω is the vibrational frequency, k is the wave vector ($k = 2\pi/\lambda$, where λ is the wavelength), related by $\omega = ck$, where c is the propagation speed, here the speed of light. It can be immediately checked that a maximum of $u(z, t)$ moves at speed $\omega/k = c$. In what follows we shall always work in a plane at fixed z, for example, the $z = 0$ plane where

$$u(z = 0, t) := u(t) = u_0 \cos \omega t.$$

When an electromagnetic wave passes through a polarizing filter (a *polarizer*), the vibration transmitted by the filter is a vector in the xOy plane transverse to the propagation direction:

$$
\begin{aligned}
E_x &= E_0 \cos\theta \cos\omega t, \\
E_y &= E_0 \sin\theta \cos\omega t,
\end{aligned}
\tag{2.1}
$$

where θ depends on the orientation of the filter. The light intensity (or energy) measured, for example, using a photoelectric cell is proportional to the squared electric field $I \propto E_0^2$ (in general, the energy of a vibration is proportional to the squared vibrational amplitude). The unit vector [1] $\hat{p}$ in the xOy plane

$$
\hat{p} = (\cos\theta, \sin\theta), \quad \vec{E} = E_0 \hat{p} \cos\omega t,
\tag{2.2}
$$

characterizes the *(linear) polarization* of the electromagnetic wave. If $\theta = 0$ the light is polarized in the x direction, and if $\theta = \pi/2$ it is polarized in the y direction. Natural light is *unpolarized* because it is made up of an *incoherent* superposition (this important concept will be defined precisely in Chapter 4) of 50% light polarized along Ox and 50% light polarized along Oy.

We shall study polarization quantitatively using a *polarizer–analyzer ensemble*. We allow the light first to pass through a polarizer whose axis makes an angle θ with Ox, and then through a second polarizer, called an analyzer, whose axis makes an angle α with Ox (Fig. 2.1), and write

$$
\hat{n} = (\cos\alpha, \sin\alpha).
\tag{2.3}
$$

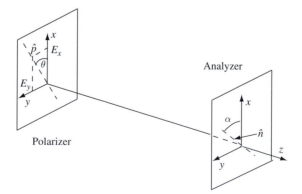

Figure 2.1 A polarizer–analyzer ensemble.

---

[1] Throughout this book, unit vectors of ordinary space $\mathbb{R}^3$ will be denoted by a hat: $\hat{p} = \vec{p}/p$, $\hat{n} = \vec{n}/n$, ...

At the exit from the analyzer the electric field $\vec{E}'$ is obtained by projecting the field (2.1) onto $\hat{n}$:

$$
\begin{aligned}
\vec{E}' = (\vec{E} \cdot \hat{n})\hat{n} &= E_0 \cos \omega t (\hat{p} \cdot \hat{n})\hat{n} \\
&= E_0 \cos \omega t (\cos \theta \cos \alpha + \sin \theta \sin \alpha)\hat{n} \qquad (2.4) \\
&= E_0 \cos \omega t \cos(\theta - \alpha)\hat{n}.
\end{aligned}
$$

From this we obtain the *Malus law* for the intensity at the exit from the analyzer:

$$
I' = I \cos^2(\theta - \alpha). \qquad (2.5)
$$

Linear polarization is not the most general possible case. *Circular polarization* is obtained by choosing $\theta = \pi/4$ and shifting the phase of the y component by $\pm\pi/2$. For example, for right-handed circular polarization we have

$$
\begin{aligned}
E_x &= \frac{E_0}{\sqrt{2}} \cos \omega t, \\
E_y &= \frac{E_0}{\sqrt{2}} \cos \left(\omega t - \frac{\pi}{2} \right) = \frac{E_0}{\sqrt{2}} \sin \omega t.
\end{aligned} \qquad (2.6)
$$

The electric field vector traces a circle of radius $|E_0|/\sqrt{2}$ in the xOy plane. The most general case is that of elliptical polarization, where the tip of the electric field vector traces an ellipse:

$$
\begin{aligned}
E_x &= E_0 \cos \theta \cos(\omega t - \delta_x) = E_0 \operatorname{Re}\left[\cos \theta \, \mathrm{e}^{-\mathrm{i}(\omega t - \delta_x)} \right] = E_0 \operatorname{Re}\left(\lambda \, \mathrm{e}^{-\mathrm{i}\omega t} \right), \\
E_y &= E_0 \sin \theta \cos(\omega t - \delta_y) = E_0 \operatorname{Re}\left[\sin \theta \, \mathrm{e}^{-\mathrm{i}(\omega t - \delta_y)} \right] = E_0 \operatorname{Re}\left(\mu \, \mathrm{e}^{-\mathrm{i}\omega t} \right).
\end{aligned} \qquad (2.7)
$$

It will be important for what follows to note that *only the difference $\delta = (\delta_y - \delta_x)$ is physically relevant*. By a simple change of time origin we can, for example, choose $\delta_x = 0$. To summarize, the most general polarization is described by a *complex* vector normalized to unity (or a *normalized vector*) in a two-dimensional space with components

$$
\lambda = \cos \theta \, \mathrm{e}^{\mathrm{i}\delta_x}, \quad \mu = \sin \theta \, \mathrm{e}^{\mathrm{i}\delta_y},
$$

and $|\lambda|^2 + |\mu|^2 = 1$. Owing to the arbitrariness in the phase, a vector with components (λ', μ'),

$$
\lambda' = \lambda \, \mathrm{e}^{\mathrm{i}\phi}, \quad \mu' = \mu \, \mathrm{e}^{\mathrm{i}\phi},
$$

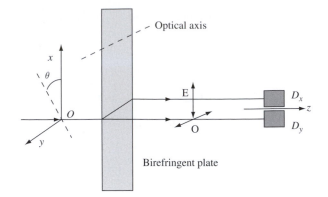

Figure 2.2 Decomposition of the polarization by a birefringent plate. The ordinary ray O is polarized horizontally and the extraordinary ray E is polarized vertically.

represents the same polarization as (λ, μ). It is more correct to say that the polarization is represented mathematically by a *ray*, that is, by a vector up to a phase.

Remarks

- A birefringent plate (Fig. 2.2) can be used to separate an incident beam into two orthogonal polarization states, and one can repeat the Malus experiment by checking that a suitably oriented polarizing filter absorbs one of the two polarizations while allowing the orthogonal one to pass through.
- Let us consider a crossed polarizer–analyzer ensemble, for example, with the polarizer aligned along Ox and the analyzer along Oy. No light is transmitted. However, if we introduce an intermediate polarizer whose axis makes an angle θ with Ox, part of the light is transmitted: the first projection gives a factor $\cos\theta$ and the second gives a factor $\sin\theta$, so that the intensity at the exit of the analyzer is

$$I' = I\cos^2\theta\sin^2\theta,$$

which vanishes only for $\theta = 0$ or $\theta = \pi/2$.

2.2 Photon polarization

Ever since the work of Einstein (1905), we have known that light is composed of photons or light particles. If the light intensity is reduced sufficiently, it should be possible to study the polarization of individual photons which can easily be detected using photodetectors, the modern version of which is the CCD (Charge Coupling Device) camera.[2] Let us suppose that $\mathcal{N}$ photons are detected in an

---

[2] A cell of the retina is sensitive to an isolated photon, but only a few percent of the photons entering the eye reach the retina.

experiment. When $\mathcal{N} \to \infty$ it should be possible to recover the results of wave optics which we have just stated above. For example, let us perform the following experiment (Fig. 2.2). A birefringent plate is used to separate a light beam whose polarization makes an angle θ with Ox into a beam polarized along Ox and a beam polarized along Oy, the intensities respectively being $I\cos^2\theta$ and $I\sin^2\theta$. We reduce the intensity such that the photons arrive one by one, and we place two photodetectors D_x and D_y behind the plate. Experiment shows that D_x and D_y are never triggered simultaneously,[3] i.e., an entire photon reaches *either* D_x or D_y: a photon is never split. On the other hand, experiment shows that the probability $\mathsf{p}_x(\mathsf{p}_y)$ that a photon is detected by $D_x(D_y)$ is $\cos^2\theta(\sin^2\theta)$. If $\mathcal{N}$ photons are detected in the experiment, we must have $\mathcal{N}_x(\mathcal{N}_y)$ photons detected by $D_x(D_y)$:

$$\mathcal{N}_x \simeq \mathcal{N}\cos^2\theta, \quad \mathcal{N}_y \simeq \mathcal{N}\sin^2\theta,$$

where $\simeq$ is used to indicate statistical fluctuations of order $\sqrt{\mathcal{N}}$. Since the light intensity is proportional to the number of photons, we recover the Malus law in the limit $\mathcal{N} \to \infty$. However, in spite of its simplicity this experiment raises two fundamental problems.

- **Problem 1** Is it possible to predict whether a given photon will trigger D_x or D_y? The response of quantum theory is NO, which profoundly shocked Einstein ("God does not play dice!"). Some physicists have tried to assume that quantum theory is incomplete, and that there are "hidden variables" whose knowledge would permit prediction of which detector a given photon reaches. If we make some very reasonable hypotheses to which we shall return in Chapter 4, we now know that such hidden variables are experimentally excluded. The probabilities of quantum theory are *intrinsic*; they are not related to imperfect knowledge of the physical situation, as is the case, for example, in the game of tossing a coin.
- **Problem 2** Let us recombine[4] the two beams from the first birefringent plate by using a second plate located symmetrically relative to the first (Fig. 2.3) and find the probability for a photon to cross the analyzer. A photon can choose path E with probability $\cos^2\theta$. Then it has probability $\cos^2\alpha$ of passing through the analyzer, or a total probability $\cos^2\theta\cos^2\alpha$. If path O is chosen, the probability of passing through the analyzer will be $\sin^2\theta\sin^2\alpha$. The total probability is obtained by adding the probabilities of the two possible paths:

$$\mathsf{p}'_{\text{tot}} = \cos^2\theta\cos^2\alpha + \sin^2\theta\sin^2\alpha. \tag{2.8}$$

---

[3] Except in the case of a "dark count," where a detector is triggered spontaneously.
[4] With some care, as the difference between the ordinary and extraordinary indices of refraction must be taken into account; *cf.* Le Bellac (2006), Exercise 3.1.

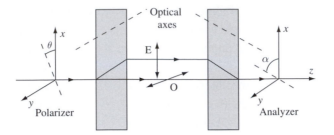

Figure 2.3 Decomposition and recombination of polarizations by means of bire-fringent plates. The photon can choose path E (extraordinary), where it is polarized along Ox, or path O (ordinary), where it is polarized along Oy.

This result is FALSE! In fact, we know from classical optics that the intensity is $I\cos^2(\theta - \alpha)$, and the correct result, confirmed by experiment, is

$$p_{\text{tot}} = \cos^2(\theta - \alpha), \tag{2.9}$$

which is not at all the same thing!

In order to recover the results of wave optics it is necessary to introduce into quantum physics the fundamental notion of a *probability amplitude* $a(\alpha \to \beta)$. A probability amplitude is a complex number, the squared modulus of which gives the probability: $p(\alpha \to \beta) = |a(\alpha \to \beta)|^2$. In the preceding example, the relevant probability amplitudes are

$$a(\theta \to x) = \cos\theta, \quad a(x \to \alpha) = \cos\alpha,$$
$$a(\theta \to y) = \sin\theta, \quad a(y \to \alpha) = \sin\alpha.$$

For example, $a(\theta \to x)$ is the probability amplitude that the photon polarized along the direction θ chooses the E path, where it is polarized along Ox. Then, a basic principle of quantum physics is that one must *add the amplitudes for indistinguishable paths*:

$$a_{\text{tot}} = \cos\theta\cos\alpha + \sin\theta\sin\alpha = \cos(\theta - \alpha),$$

which allows us to recover (2.9):

$$p_{\text{tot}} = |a_{\text{tot}}|^2 = \cos^2(\theta - \alpha).$$

The superposition of probability amplitudes in a_{tot} is the exact analog of the superposition of wave amplitudes: the laws for combining quantum amplitudes are exact copies of those of wave optics, and the results of the latter are recovered in the limit of a large number of photons. Let us suppose, however, that we have some way of knowing whether a photon has followed path E or path O (this is impossible in our case, but similar experiments to determine the path, termed "which path experiments," have been performed using atoms). We can then divide the photons into two classes, those which have chosen path E and those which have chosen path O. For the former we could have blocked path O by a mask without changing anything, and the reverse for the latter photons. The result can obviously only be (2.8). If we manage to distinguish between the paths, the result will no longer be (2.9), because the paths are no longer indistinguishable!

Under experimental conditions where it is impossible in principle to distinguish between the paths, we can make one or the other statement:

- the photon is able to explore both paths at the same time, or
- (the author's preference) it makes no sense to ask the question "Which path?", because the experimental conditions do not permit it to be answered. We shall follow Asher Peres, who states "Unperformed experiments have no results!"

It should be noted that if the experiment allows us to distinguish between the two paths, the result is (2.8), even if we decide not to observe which path is followed. It is sufficient that the experimental conditions *in principle* allow the two paths to be distinguished, even when the current technology does not permit this to be done in practice.

We have examined a particular case of a quantum phenomenon, the photon polarization, but the results we have described have led us to the very core of quantum physics.

2.3 Mathematical formulation: the qubit

The photon polarization can be used to transmit information, for example, by an optical fiber. We can arbitrarily decide to associate the bit value 0 with a photon polarized along Ox and the bit value 1 with a photon polarized along Oy. In quantum information theory the people who exchange information are conventionally called Alice (A) and Bob (B). For example, Alice sends Bob a series of photons polarized as

$$yyxyxyyyx\cdots.$$

Bob analyzes the polarization of these photons using a birefringent plate as in Fig. 1.2 and deciphers the message sent by Alice:

$$110101110\cdots.$$

This is obviously not a very efficient way of exchanging messages. However, we shall see that this protocol forms the basis of quantum cryptography. An interesting question now is, what bit value can be associated with, for example, a photon polarized at 45°? According to the results of the preceding section, a

photon polarized at 45° is a *linear superposition* of a photon polarized along
Ox and a photon polarized along *Oy*. The photon polarization gives an example
of a qubit, and a qubit is therefore a much richer object than an ordinary bit,
which can take only the values 0 and 1. In a certain sense, a qubit can take all
values intermediate between 0 and 1 and therefore contains an infinite amount
of information! However, this optimistic statement is immediately deflated when
we recall that measurement of a qubit can give only the result 0 or 1, no matter
which basis is chosen: a photon either chooses the E path (value 0 of the bit)
or the O path (value 1 of the bit) and this result holds whatever the orientation
of the birefringent plate. Nevertheless, we can ask the question whether or not
this "hidden information" contained in the linear superposition is valuable, and in
Chapter 5 we shall see that under certain conditions this information can actually
be exploited.

In order to take into account linear superpositions, it is natural to introduce a
two-dimensional vector space $\mathcal{H}$ for the mathematical description of polarization.
Any polarization state can be put into correspondence with a vector in this space.
We can, for example, choose as orthogonal basis vectors of $\mathcal{H}$ the vectors $|x\rangle$
and $|y\rangle$ corresponding to linear polarizations along *Ox* and *Oy*. Any polarization
state can be decomposed on this basis: [5]

$$|\Phi\rangle = \lambda|x\rangle + \mu|y\rangle. \tag{2.10}$$

We use the Dirac notation for the vectors of $\mathcal{H}$; see Box 2.1. There exists a
very precise experimental procedure for constructing the state $|\Phi\rangle$; it is described
in detail in Exercise 2.6.2. A linear polarization will be described using real
coefficients λ and μ, but the description of a circular (2.6) or elliptical (2.7)
polarization will require coefficients λ and μ which are complex. The space $\mathcal{H}$
is therefore a *complex vector space*, isomorphic to $\mathbb{C}^2$.

Probability amplitudes are associated with scalar products on this space. Let us
take two vectors, $|\Phi\rangle$ given by (2.10) and $|\Psi\rangle$ given by

$$|\Psi\rangle = \nu|x\rangle + \sigma|y\rangle.$$

The *scalar product* of these vectors will be denoted $\langle\Psi|\Phi\rangle$, and by definition

$$\langle\Psi|\Phi\rangle = \nu^*\lambda + \sigma^*\mu = \langle\Phi|\Psi\rangle^*, \tag{2.11}$$

where c^* is the complex conjugate of c. This scalar product is therefore linear in
$|\Phi\rangle$ and antilinear in $|\Psi\rangle$. It defines the *norm* $||\Phi||$ of the vector $|\Phi\rangle$:

$$||\Phi||^2 = \langle\Phi|\Phi\rangle = |\lambda|^2 + |\mu|^2. \tag{2.12}$$

---

[5] We use upper-case Greek letters for generic vectors of $\mathcal{H}$ in order to avoid confusion with the vectors
representing linear polarizations such as $|\theta\rangle$, $|\alpha\rangle$, etc.

Box 2.1: Dirac notation

"Mathematicians tend to loathe the Dirac notation, because it prevents them from making distinctions they consider important. Physicists love the Dirac notation because they are always forgetting that such distinctions exist and the notation liberates them from having to remember" (Mermin (2003)). In our presentation here the Dirac notation reduces to a simple notational convention and avoids matters of principle.

Let $\mathcal{H}^{(N)}$ be a Hilbert space of finite dimension N on the complex numbers and let u, v, w be vectors of $\mathcal{H}^{(N)}$. The scalar product of two vectors v and w is denoted (v, w), following for the time being the mathematicians' notation. It satisfies [6]

$$(v, \lambda w + \mu w) = \lambda(v, w) + \mu(v, w'), \quad (v, w) = (w, v)^*.$$

Let $\{e_n\}$ be an orthonormal basis of $\mathcal{H}^{(N)}$, $n = 1, 2, \ldots, N$. In this basis the vectors (u, v, w) have the components

$$u_n = (e_n, u), \quad v_n = (e_n, v), \quad w_n = (e_n, w).$$

Let us consider a linear operator $A(v, w)$ defined by its matrix representation in the basis $\{e_n\}$:

$$A_{nm}(v, w) = v_n w_m^*.$$

The action of this operator on the vector u, $u \xrightarrow{A} u'$, is given in terms of components by

$$u'_n = \sum_m A_{nm}(v, w)u_m = \sum_m v_n w_m^* u_m = \sum_m v_n (w_m^* u_m) = v_n(w, u),$$

or in vector form

$$u' = A(v, w)u = v(w, u).$$

In Dirac notation, vectors are written as $|v\rangle$ and scalar products as $\langle w|v\rangle$:

$$v \rightarrow |v\rangle, \quad (w, v) \rightarrow \langle w|v\rangle.$$

With this notation the action of $A(v, w)$ is written as

$$|u'\rangle = |A(v, w)u\rangle = |v\rangle\langle w|u\rangle$$
$$= (|v\rangle\langle w|)|u\rangle,$$

and the second line of this equation suggests the *notational convention*

$$A(v, w) = |v\rangle\langle w|.$$

---

[6] The convention of physicists differs from that of mathematicians in that for the latter the scalar product is antilinear in the second vector:

$$(v, \lambda w + \mu w') = \lambda^*(v, w) + \mu^*(v, w').$$

A case of particular importance is that where $v = w$ and v is a normalized vector. Then

$$A(v, v) = |v\rangle\langle v|, \quad |A(v, v)u\rangle = |v\rangle\langle v|u\rangle,$$

and $A(v, v)$ is the *projector* $\mathcal{P}_v$ onto the vector v, because $\langle v|u\rangle$ is the component of u along v. A familiar example is the projection in $\mathbb{R}^3$ of a vector $\vec{u}$ onto a unit vector $\hat{v}$:

$$\mathcal{P}_{\hat{v}}\vec{u} = \hat{v}(\vec{u} \cdot \hat{v}).$$

It is customary to use $|n\rangle$ to denote the vectors of an orthonormal basis: $e_n \to |n\rangle$, and the projector onto $|n\rangle$ then is

$$\mathcal{P}_n = |n\rangle\langle n|.$$

Let $\mathcal{H}^{(M)}$ be an M-dimensional ($M \leq N$) subspace of $\mathcal{H}^{(N)}$, and $|m\rangle$, $m = 1, 2, \ldots, M$ be an orthonormal basis in this subspace. The projector onto $\mathcal{H}^{(M)}$ then is

$$\mathcal{P}_{\mathcal{H}^{(M)}} = \sum_{m=1}^{M} |m\rangle\langle m|,$$

and if $M = N$ we obtain the decomposition of the identity, which physicists call the *completeness relation*:

$$\sum_{m=1}^{N} |m\rangle\langle m| = I,$$

where I is the identity operator. The matrix elements of a linear operator A are given by

$$A_{mn} = \langle m|An\rangle$$

and the completeness relation can be used, for example, to find immediately the matrix multiplication law:

$$(AB)_{mn} = \langle m|ABn\rangle = \langle m|AIBn\rangle = \sum_k \langle m|Ak\rangle\langle k|Bn\rangle = \sum_k A_{mk}B_{kn}.$$

The vectors $|x\rangle$ and $|y\rangle$ are orthogonal with respect to the scalar product (2.11) and they have unit norm:

$$\langle x|x\rangle = \langle y|y\rangle = 1, \qquad \langle x|y\rangle = 0.$$

The basis $\{|x\rangle, |y\rangle\}$ is therefore an *orthonormal basis* of $\mathcal{H}$. To the definition of a physical state we shall add the convenient, but not essential, normalization condition

$$||\Phi||^2 = |\lambda|^2 + |\mu|^2 = 1. \tag{2.13}$$

Polarization states will therefore be represented mathematically by normalized vectors (vectors of unit norm) in the space $\mathcal{H}$; they are called *state vectors* (of polarization). A vector space on which a positive-definite scalar product is defined is called a *Hilbert space*, and $\mathcal{H}$ is the *Hilbert space (of polarization states)*.

Now let us return to the probability amplitudes. A state linearly polarized along θ will be denoted $|\theta\rangle$ with

$$|\theta\rangle = \cos\theta\,|x\rangle + \sin\theta|y\rangle. \tag{2.14}$$

The vector $|\theta\rangle$ gives the mathematical description of the linear polarization state of a photon. The probability amplitude for a photon polarized along θ to pass through an analyzer oriented along α is, as we have seen above,

$$a(\theta \to \alpha) = \cos(\theta - \alpha) = \langle\alpha|\theta\rangle. \tag{2.15}$$

It is therefore given by the scalar product of the vectors $|\alpha\rangle$ and $|\theta\rangle$, and the probability of passing through the analyzer is given by the squared modulus of this amplitude (see (2.9)):

$$\mathsf{p}(\theta \to \alpha) = \cos^2(\theta - \alpha) = |\langle\alpha|\theta\rangle|^2. \tag{2.16}$$

A probability amplitude ("the amplitude of the probability for finding $|\Phi\rangle$ in $|\Psi\rangle$," where $|\Phi\rangle$ and $|\Psi\rangle$ represent general polarization states) will be defined in general as

$$a(\Phi \to \Psi) = \langle\Psi|\Phi\rangle, \tag{2.17}$$

and the corresponding probability will be given by

$$\mathsf{p}(\Phi \to \Psi) = |a(\Phi \to \Psi)|^2 = |\langle\Psi|\Phi\rangle|^2. \tag{2.18}$$

It is important to note that a state vector is actually defined only up to a multiplicative phase; for example, in (2.10) we can multiply λ and μ by the *same* phase factor

$$(\lambda, \mu) \equiv (e^{i\delta}\lambda, e^{i\delta}\mu),$$

because replacing $|\Phi\rangle$ by

$$|\Phi'\rangle = e^{i\delta}|\Phi\rangle$$

leaves the probabilities $|\langle\Psi|\Phi\rangle|^2$ unchanged whatever $|\Psi\rangle$, and these probabilities are the only quantities which can be measured. A multiplicative global phase is not physically relevant; the correspondence is therefore not between a physical state and a vector, but rather between a physical state and a *ray*, that is, a vector up to a phase.

Now we are ready to tackle the crucial question of *measurement* in quantum physics. Measurement is based on two notions, that of the preparation of a quantum state and that of a test. We again use the polarizer–analyzer ensemble and assume that the analyzer, which prepares the polarization state, is oriented along Ox. If the polarizer is also oriented along Ox, a photon leaving the polarizer passes through the analyzer with 100% probability, while if the polarizer is oriented along Oy, the probability is zero. The analyzer performs a *test* (of the polarization), and the result of the test is 1 or 0. The test then allows the polarization state of the photon to be determined. However, this is not the case in general. Let us assume that the polarizer is oriented in the direction θ or in the orthogonal direction $\theta_\perp$:

$$|\theta\rangle = \cos\theta\,|x\rangle + \sin\theta\,|y\rangle,$$
$$|\theta_\perp\rangle = -\sin\theta\,|x\rangle + \cos\theta\,|y\rangle. \tag{2.19}$$

The states $|\theta\rangle$ and $|\theta_\perp\rangle$, like the states $|x\rangle$ and $|y\rangle$, form an orthonormal basis of $\mathcal{H}$. If, for example, the polarizer prepares the photon in the state $|\theta\rangle$ and the analyzer is oriented along Ox, then the probability of passing the test is $\cos^2\theta$. Two essential things should be noted:

- After the passage through the analyzer, the polarization state of the photon is no longer $|\theta\rangle$, but $|x\rangle$. It is often said that *the measurement perturbs the polarization state*. However, this statement is debatable: the measurement performed by the analyzer is a measurement of the physical property "polarization of the photon along Ox," but this polarization does not exist before the measurement because the photon is in the state $|\theta\rangle$, and that which does not exist cannot be perturbed! We shall illustrate this by another example at the end of this section.
- If the photon is elliptically, rather than linearly, polarized,

$$\lambda = \cos\theta, \quad \mu = \sin\theta\,\mathrm{e}^{\mathrm{i}\delta}, \quad \delta \neq 0,$$

the probability of passing the test is again $\cos^2\theta$: the test does not permit an unambiguous determination of the polarization. *Only if the probability of passing the test is 0 or 1 does the measurement permit the unambiguous determination of the initial polarization state. Therefore, unless one knows beforehand the basis in which it has been prepared, there is no test which permits the unambiguous determination of the polarization state of an isolated photon.* As explained in Exercise 2.6.1, determination of the polarization of a light wave, or of a large number of identically prepared photons, is possible provided one uses two different orientations of the analyzer.

There is thus a difference of principle between a measurement in classical physics and one in quantum physics. In classical physics *the physical quantity which is measured exists before the measurement*: if radar is used to measure the speed of your car equal to 180 km/h on the highway, this speed existed before the police performed the measurement (thus giving them the right to issue a speeding ticket). On the contrary, in the measurement of the photon polarization $|\theta\rangle$ by an analyzer oriented along Ox, the fact that the test gives a polarization along Ox does not permit us to conclude that the tested photon actually had polarization along Ox before the measurement. Again taking the analogy to a car, we can imagine that as in (2.19) the car is in a linear superposition of two speed states, [7] for example,

$$|v\rangle = \sqrt{\frac{1}{3}}|120\,\mathrm{km/h}\rangle + \sqrt{\frac{2}{3}}|180\,\mathrm{km/h}\rangle.$$

The police will measure a speed of 120 km/h with probability 1/3 and a speed of 180 km/h with probability 2/3, but it would be incorrect to think that one of the two results existed before the measurement. Quantum logic is incompatible with classical logic!

2.4 Principles of quantum mechanics

The principles of quantum mechanics generalize the results we have obtained in the case of photon polarization.

- **Principle 1** The physical state of a quantum system is represented by a vector $|\Phi\rangle$ belonging to a Hilbert space $\mathcal{H}$ of, in general, infinite dimension. Fortunately, for the purposes of quantum information theory, we only need spaces of finite dimension. Unless explicitly stated otherwise, $|\Phi\rangle$ will be chosen to be a normalized vector: $||\Phi||^2 = 1$. $|\Phi\rangle$ is called the *state vector* of the quantum system.
- **Principle 2** If $|\Phi\rangle$ and $|\Psi\rangle$ represent two physical states, the probability amplitude $a(\Phi \to \Psi)$ of finding Φ in Ψ is given by the scalar product $\langle\Psi|\Phi\rangle$:

$$a(\Phi \to \Psi) = \langle\Psi|\Phi\rangle,$$

and the probability for Φ to pass the Ψ test is

$$p(\Phi \to \Psi) = |a(\Phi \to \Psi)|^2 = |\langle\Psi|\Phi\rangle|^2.$$

We perform this test by first using a device to prepare the quantum system in the state $|\Phi\rangle$ (a polarizer), and then using as an analyzer a second device which would have prepared the system in the state $|\Psi\rangle$.

---

[7] Of course, no one knows how to realize such a superposition state for a car, but we do know very well how to construct a superposition of states with different speeds for an elementary particle or an atom.

After the test the quantum system is in the state $|\Psi\rangle$, which from the mathematical point of view means that we have performed an orthogonal projection onto $|\Psi\rangle$. Let $\mathcal{P}_\Psi$ be the projector. Since [8]

$$|\mathcal{P}_\Psi \Phi\rangle \equiv \mathcal{P}_\Psi |\Phi\rangle = |\Psi\rangle\langle\Psi|\Phi\rangle = (|\Psi\rangle\langle\Psi|)|\Phi\rangle,$$

this projector can be written in the very convenient form (see Box 2.1)

$$\mathcal{P}_\Psi = |\Psi\rangle\langle\Psi|. \tag{2.20}$$

In summary, the mathematical operation corresponding to a measurement is a projection, and the corresponding measurement is called a *projective measurement*. However, the vector $\mathcal{P}_\Psi|\Phi\rangle$ is not in general normalized. We must then normalize it

$$\mathcal{P}_\Psi|\Phi\rangle \rightarrow |\Phi'\rangle = \frac{\mathcal{P}_\Psi|\Phi\rangle}{\langle\Phi|\mathcal{P}_\Psi\Phi\rangle}.$$

In the orthodox interpretation of quantum mechanics, the projection of a state vector followed by its normalization is called "state-vector collapse" or, for historical reasons, "wave-packet collapse." The idea of state-vector collapse is a convenient fiction of the orthodox interpretation which avoids having to ask questions about the measurement process, and it is often treated as a supplementary basic principle of quantum mechanics. However, we can perfectly well bypass this principle if we take into account the full complexity of the measurement process. An example will be given in Chapter 5, Box 5.2.

Let us now turn to the mathematical description of the physical properties of a quantum system, first by returning to polarization. In the basis $\{|x\rangle, |y\rangle\}$ the projectors $\mathcal{P}_x$ and $\mathcal{P}_y$ onto these basis states are

$$\mathcal{P}_x = |x\rangle\langle x| = \begin{pmatrix} 1 & 0 \\ 0 & 0 \end{pmatrix}, \quad \mathcal{P}_y = |y\rangle\langle y| = \begin{pmatrix} 0 & 0 \\ 0 & 1 \end{pmatrix}.$$

We note that the identity operator I can be written as the sum of the two projectors $\mathcal{P}_x$ and $\mathcal{P}_y$:

$$\mathcal{P}_x + \mathcal{P}_y = |x\rangle\langle x| + |y\rangle\langle y| = I.$$

This is a special case of the *completeness relation* (Box 2.1), which can be generalized to an orthonormal basis of a Hilbert space $\mathcal{H}$ of dimension N:

$$\sum_{i=1}^{N} |i\rangle\langle i| = I, \quad \langle i|j\rangle = \delta_{ij}.$$

---

[8] The action of an operator M on a vector $|\Phi\rangle$ will be written either as $M|\Phi\rangle$ or as $|M\Phi\rangle$.

The projectors $\mathcal{P}_x$ and $\mathcal{P}_y$ commute:

$$[\mathcal{P}_x, \mathcal{P}_y] \equiv \mathcal{P}_x \mathcal{P}_y - \mathcal{P}_y \mathcal{P}_x = 0,$$

where we denote $[A, B] := AB - BA$ the *commutator* of two operators A and B. The *tests* $|x\rangle$ and $|y\rangle$ are termed *compatible*. On the contrary, the projectors onto the states $|\theta\rangle$ and $|\theta_\perp\rangle$ (2.19),

$$\mathcal{P}_\theta = |\theta\rangle\langle\theta| = \begin{pmatrix} \cos^2\theta & \sin\theta\cos\theta \\ \sin\theta\cos\theta & \sin^2\theta \end{pmatrix},$$

$$\mathcal{P}_{\theta_\perp} = |\theta_\perp\rangle\langle\theta_\perp| = \begin{pmatrix} \sin^2\theta & -\sin\theta\cos\theta \\ -\sin\theta\cos\theta & \cos^2\theta \end{pmatrix},$$

do not commute with $\mathcal{P}_x$ and $\mathcal{P}_y$, as can be verified immediately by explicit calculation:

$$[\mathcal{P}_x, \mathcal{P}_\theta] = \begin{pmatrix} 0 & \sin\theta\cos\theta \\ -\sin\theta\cos\theta & 0 \end{pmatrix}.$$

The *tests* $|x\rangle$ and $|\theta\rangle$ are termed *incompatible*. The projectors $\mathcal{P}_x, \ldots, \mathcal{P}_{\theta_\perp}$ represent mathematically the physical properties of the quantum system when the photon is polarized along the $x, \ldots, \theta_\perp$ axes. *It is not possible to measure incompatible properties of a quantum system simultaneously.*

In the general case of a Hilbert space of states $\mathcal{H}^{(N)}$ with dimension N, to an orthonormal basis $|n\rangle$, $n = 1, \ldots, N$, of this space will be associated a set of N compatible tests $|n\rangle$. If the quantum system is in a state $|\Phi\rangle$, the probability that it passes the test is $\mathsf{p}_n = \langle n|\Phi\rangle|^2$ (**Principle 2**) and $\sum_n \mathsf{p}_n = 1$. The tests $|n\rangle$ form a *maximal test*. To a different orthonormal basis of $\mathcal{H}^{(N)}$ will correspond another maximal test incompatible with the preceding one. Now, one may ask the following question: can one define in a Hilbert space $\mathcal{H}^{(N)}$ bases $\{|n\rangle\}$ and $\{|\alpha\rangle\}$, where $n, \alpha = 1, 2, \ldots, N$, which are maximally incompatible? The answer to this question is positive. Two bases are maximally incompatible if they are *complementary*, which means by definition that $|\langle\alpha|n\rangle|^2$ is independent of n and α

$$|\langle\alpha|n\rangle|^2 = \frac{1}{N}. \tag{2.21}$$

For example, the bases $\{|x\rangle, |y\rangle\}$ and $\{|\theta = \pi/4\rangle, |\theta = -\pi/4\rangle\}$ are complementary. Any linear polarization basis is complementary to a circular polarization basis $\{|R\rangle, |L\rangle\}$ defined in Exercise 2.6.3. One way to obtain complementary bases in $\mathcal{H}^{(N)}$ is to use discrete Fourier transforms (see Section 5.7)

$$|\alpha\rangle = \frac{1}{\sqrt{N}} \sum_{n=1}^{N} e^{2i\pi\alpha n}|n\rangle.$$

Let us explain the physical meaning of complementary bases with the following example: suppose you want to test a large number of quantum systems all prepared in a state of the basis $\{|n\rangle\}$, but you do not know which one. If you test the system using the basis $\{|n\rangle\}$, one of the results, say m, will come out with 100% probability, so that your measurement gives you maximum knowledge of the state. If, on the contrary, you test the preparation using the basis $\{|\alpha\rangle\}$, then you will get all the possible outcomes with probability $1/N$, and you will get minimum knowledge of the preparation. The concept of complementary bases will be very useful for understanding the principles of quantum cryptography.

For later developments it will be useful to note that knowledge of the probabilities of passing a test $\mathcal{T}$ permits definition of the *expectation value* $\langle \mathcal{T} \rangle$:

$$\langle \mathcal{T} \rangle = 1 \times \mathsf{p}(\mathcal{T} = 1) + 0 \times \mathsf{p}(\mathcal{T} = 0) \quad [= \mathsf{p}(\mathcal{T} = 1)].$$

For example, if the test $\mathcal{T}$ is represented by the procedure $|\Psi\rangle$ and it is applied to a state $|\Phi\rangle$, then

$$\mathsf{p}(\Psi) = |\langle \Psi | \Phi \rangle|^2 = \langle \Phi | \Psi \rangle \langle \Psi | \Phi \rangle = \langle \Phi | (|\Psi\rangle \langle \Psi|) | \Phi \rangle = \langle \Phi | \mathcal{P}_\Psi \Phi \rangle. \quad (2.22)$$

In quantum physics it is standard to refer to the quantity

$$\boxed{\langle \Phi | M \Phi \rangle \equiv \langle M \rangle_\Phi} \quad (2.23)$$

as the *expectation value of the operator M in the state* $|\Phi\rangle$. The test $\mathcal{T} = |\Psi\rangle$ can therefore be associated with a projector $\mathcal{P}_\Psi$ whose expectation value in the state $|\Phi\rangle$ gives, according to (2.22), the probability of passing the test.

The generalization of this observation permits us to construct the physical properties of a quantum system using projectors. Let us give an example, again from the case of polarization. We assume that we have constructed (in a completely arbitrary way) a physical property $\mathcal{M}$ of a photon as follows: $\mathcal{M}$ is $+1$ if the photon is polarized along Ox and $\mathcal{M}$ is -1 if the photon is polarized along Oy. With the physical property $\mathcal{M}$ we can associate a Hermitian operator M,

$$M = \mathcal{P}_x - \mathcal{P}_y,$$

satisfying

$$M|x\rangle = +|x\rangle, \quad M|y\rangle = -|y\rangle.$$

The expectation value of M is, by definition,

$$\langle M \rangle = 1 \times \mathsf{p}(M = 1) + (-1) \times \mathsf{p}(M = -1).$$

Let us assume that the photon is in the state $|\theta\rangle$; then the expectation value $\langle M \rangle_\theta$ in the state $|\theta\rangle$ is

$$\langle M \rangle_\theta = \langle \theta | \mathcal{P}_x \theta \rangle - \langle \theta | \mathcal{P}_y \theta \rangle = \cos^2 \theta - \sin^2 \theta = \cos 2\theta.$$

The operator M thus constructed is a Hermitian operator ($M = M^\dagger$ or $M_{ij} = M_{ji}^*$), and in general *physical properties* in quantum mechanics are represented mathematically by Hermitian operators, often called *observables*. We have constructed M starting from projectors, but reciprocally we can construct projectors starting from a Hermitian operator M owing to the *spectral decomposition theorem*, which we state without proof.

Theorem Let M be a Hermitian operator. Then M can be written as a function of a set of projectors $\mathcal{P}_n$ satisfying

$$M = \sum_n a_n \mathcal{P}_n, \tag{2.24}$$

$$\mathcal{P}_n \mathcal{P}_m = \mathcal{P}_n \delta_{mn}, \qquad \sum_n \mathcal{P}_n = I, \tag{2.25}$$

where the real coefficients a_n are the eigenvalues of M. The projectors $\mathcal{P}_n$ are orthogonal to each other (but in general they project onto a subspace of $\mathcal{H}$ and not onto a single vector of $\mathcal{H}$), and their sum is the identity operator.

Let us summarize the results on the physical properties. The physical properties of a quantum system are represented mathematically by Hermitian operators and the measurement of a physical property $\mathcal{M}$ has as its result one of the eigenvalues a_n of the operator M

$$M|n\rangle = a_n |n\rangle.$$

In order to simplify the discussion, we assume that the eigenvalues of M are nondegenerate, so that the spectral decomposition (2.24) and (2.25) becomes

$$M = \sum_{n=1}^N a_n |n\rangle \langle n| \quad I = \sum_{n=1}^N |n\rangle \langle n|, \tag{2.26}$$

where N is the dimension of the Hilbert space of states. If the quantum system is in the eigenstate $|n\rangle$, the value taken by $\mathcal{M}$ is *exactly* a_n. If the quantum system is in the state $|\Phi\rangle$, then the probability of finding it in $|n\rangle$ is, from (2.18)

$$\mathsf{p}_n = |\langle n|\Phi\rangle|^2 = \langle \Phi|n\rangle \langle n|\Phi\rangle.$$

If one measures the value a_n of M, then the state vector after measurement is $|n\rangle$: this is the state vector collapse. The expectation value of $\mathcal{M}$ is by definition

$$\sum_{n=1}^{N} \mathsf{p}_n a_n = \sum_{n=1}^{N} \langle\Phi|n\rangle a_n\langle n|\Phi\rangle = \langle\Phi|M\Phi\rangle = \langle M\rangle_\Phi, \qquad (2.27)$$

which justifies the definition (2.23). This expectation value has the following physical interpretation: in an experiment conducted with a large number $\mathcal{N}$ of quantum systems all prepared in the same state $|\Phi\rangle$, the average value of the measurements of M is $\langle M\rangle_\Phi$

$$\langle M\rangle_\Phi = \lim_{\mathcal{N}\to\infty} \frac{1}{\mathcal{N}}(M_1 + \cdots + M_\mathcal{N}) \qquad (2.28)$$

where M_i is the result of the measurement number i, which is necessarily one of the eigenvalues a_n of M. We leave to the reader the generalization of the preceding results to the case where M has degenerate eigenvalues.

Box 2.2: A quantum random-number generator

It is often necessary to generate random numbers, for example, for use in Monte Carlo simulations. All computers contain a program for random number generation. However, these numbers are generated by an algorithm, and they are not actually random, but only *pseudo-random*. A simple algorithm (too simple to be reliable!) consists of, for example, calculating

$$I_{n+1} \equiv aI_n + b \bmod \mathrm{M}, \qquad 0 \le I_n \le M - 1,$$

where a and b are given integers and M is an integer, $M \gg 1$. The series $I'_n = I_n/M$ is a series of pseudo-random numbers in the interval $[0, 1]$. In some cases the inevitable regularities in a series of pseudo-random numbers can lead to errors in numerical simulations. Quantum properties can be used experimentally to realize generators of numbers which are truly random and not pseudo-random; as we shall see in the following section, truly random numbers are essential for quantum cryptography. One of the simplest devices uses a semi-transparent plate or beam-splitter. If a light ray falls on a beam-splitter, part of the light is transmitted and part is reflected. This can be arranged such that the proportions are 50%/50%. If the intensity is then decreased such that the photons arrive one by one at the plate, these photons can be either reflected and detected by D_1, or transmitted and detected by D_2 (Fig. 2.4). There is no correlation between the detections, and so this amounts to a true, unbiased coin toss. A prototype based on this principle has been realized by the quantum optics group in Geneva. It generates random numbers at a rate of 10^5 numbers per second, and the absence of any bias (equivalently, correlations between numbers supposed to be random) has been verified using standard programs.

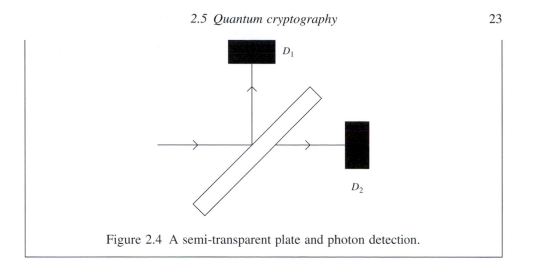

Figure 2.4 A semi-transparent plate and photon detection.

2.5 Quantum cryptography

Quantum cryptography is a recent invention based on the incompatibility of two different bases of linear polarization states. Ordinary cryptography is based on an encryption key known only to the sender and receiver and is called *secret-key cryptography*. It is in principle very secure, [9] but the sender and receiver must have a way of exchanging the key without it being intercepted by a spy. The key must be changed frequently, because a set of messages encoded with the same key can reveal regularities which permit decipherment by a third person. The transmission of a secret key is a risky process, and for this reason it is now preferred to use systems based on a different principle, the so-called *public-key* systems. In these the key is announced publicly, for example, via the Internet. A public-key system currently in use [10] is based on the difficulty of factoring a very large number N into primes, whereas the reverse operation can be done immediately: even without the help of a pocket calculator one can find $137 \times 53 = 7261$ in a few seconds, but given 7261 it would take a some time to factor it into primes. Using the best current algorithms, the time needed for a computer to factor a number N into primes grows with N as $\simeq \exp[1.9(\ln N)^{1/3}(\ln\ln N)^{2/3}]$. The current record is 176 digits, and it takes several months for a PC cluster to factorize such a number. In public-key encryption the receiver, conventionally named Bob, publicly announces to the sender, conventionally named Alice, a very large number $N = pq$ which is the product of two prime numbers p and q, along with another number c (see Box 2.3). These two numbers N and c are sufficient for Alice to encode the message, but the numbers p and q are needed to decipher it. Of course, a spy

---

[9] An absolutely secure encryption was discovered by Vernam in 1917. However, absolute security requires that the key be as long as the message and that it be used only a single time!

[10] Called RSA, as it was invented by Rivest, Shamir, and Adleman in 1977.

Box 2.3: RSA encryption (see also Box 5.3)

Bob chooses two primes p and q, $N = pq$, and a number c having no common divisor with the product $(p-1)(q-1)$. He calculates d, the inverse of c for mod $(p-1)(q-1)$ multiplication:

$$cd \equiv 1 \bmod (p-1)(q-1).$$

By a non-secure path he sends Alice the numbers N and c (but not p and q separately!). Alice wants to send Bob an encoded message, which must be represented by a number $a < N$ (if the message is too long, Alice can split it into several sub-messages). She then calculates (Fig. 2.5)

$$b \equiv a^c \bmod N$$

and sends b to Bob, always by a non-secure path, because a spy who knows only N, c, and b cannot deduce the original message a. When Bob receives the message he calculates

$$b^d \bmod N = a.$$

The fact that the result is precisely a, that is, the original message of Alice, is a result from number theory (see Box 5.3 for a proof of this result). To summarize, the numbers N, c, and b are sent publicly, by a non-secure path.

Example

$$p = 3, \quad q = 7, \quad N = 21, \quad (p-1)(q-1) = 12.$$

The number $c = 5$ has no common factor with 12, and its inverse with respect to mod 12 multiplication is $d = 5$ because $5 \times 5 = 24 + 1$. Alice chooses $a = 4$ for her message. She calculates

$$4^5 = 1024 = 21 \times 48 + 16, \quad 4^5 \equiv 16 \bmod 21.$$

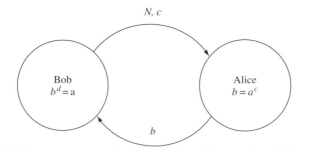

Figure 2.5 RSA encryption scheme. Bob chooses $N = pq$ and c. Alice encrypts her message a using $b = a^c$ and Bob decrypts it using $b^d = a$.

Alice then sends Bob the message 16. Bob calculates

$$b^5 = 16^5 = 49.932 \times 21 + 4, \quad 16^5 \equiv 4 \bmod 21,$$

thus recovering the original message $a = 4$. The above calculation of $16^5 \bmod 21$, for example, has not been done very cleverly. Instead, we can calculate $16^2 \bmod 21 = 4$, and then $16^3 \bmod 21$ as $4 \times 16 \bmod 21 = 1$, from which without further calculation we find $16^5 \bmod 21 = 4$. This method can be used to manipulate only numbers which are not very large compared to N.

(conventionally named Eve) possessing a sufficiently powerful computer and enough time will eventually crack the code, but one can in general count on the message being kept secret for a limited period of time. However, it is not impossible that one day we will possess very powerful algorithms for decomposing a number into primes, and moreover, if quantum computers ever see the light of day such factorization will become quite simple, at least in principle. Happily, thanks to quantum mechanics we are nearly at the point of being able to thwart the efforts of spies!

"Quantum cryptography" is a catchy phrase, but it is somewhat inaccurate. A better terminology is *quantum key distribution* (QKD). In fact, there is no encryption of a message using quantum physics; the latter is used only to ensure that the transmission of a key is not intercepted by a spy. As we have already explained, a message, encrypted or not, can be transmitted using the two orthogonal linear polarization states of a photon, for example, $|x\rangle$ and $|y\rangle$. We can choose to associate the value 0 with the polarization $|x\rangle$ and the value 1 with the polarization $|y\rangle$, so that each photon will carry a bit of information. Any message, encrypted or not, can be written in binary language as a series of 0s and 1s. The message 0110001 will be encoded by Alice by the photon sequence $xyyxxxy$, which she will send to Bob via, for example, an optical fiber. Bob will use a birefringent plate to separate the photons of vertical and horizontal polarization as in Fig. 2.2, and two detectors placed behind the plate will tell him whether the photon was polarized horizontally or vertically, so that he can reconstruct the message. If the message were just an ordinary one, there would certainly be much easier and more efficient ways of sending it! Let us simply note that if Eve taps into the fiber, detects the photons, and then resends to Bob photons of polarization identical to the ones sent by Alice, then Bob has no way of knowing that the transmission has been intercepted. The same would be true for any apparatus functioning in a classical manner (that is, not using the superposition principle): if the spy takes sufficient precautions, the spying is undetectable.

This is where quantum mechanics and the superposition principle come to the aid of Alice and Bob, by allowing them to be sure that their message has not been intercepted. The message need not be long (the transmission scheme based on polarization is not very efficient). In general, one wishes to transmit a key which permits the encryption of a later message, a key which can be replaced whenever desired. Alice sends Bob photons of four types, polarized along $Ox(\updownarrow)$ and $Oy(\longleftrightarrow)$, as before, and polarized along axes rotated by $\pm 45°$: $Ox'(\diagdown)$ and $Oy'(\diagup)$, respectively corresponding to bit values 0 and 1 (Fig. 2.6). Note that the two bases are complementary, or maximally incompatible. Similarly, Bob analyzes the photons sent by Alice using analyzers which can be oriented in four directions, vertical/horizontal and $\pm 45°$. One possibility is to use a birefringent crystal randomly oriented either vertically or at 45° with respect to the vertical and to detect the photons leaving the crystal as in Fig. 2.3. However, instead of rotating the crystal+detectors ensemble, it is easier to use a Pockels cell, which allows a given polarization to be transformed into an arbitrary polarization while maintaining the crystal+detectors ensemble fixed. An example is given in Fig. 2.7. Bob records a 0 if the photon has polarization $\updownarrow$ or $\diagup$ and 1 if it has polarization $\longleftrightarrow$ or $\diagup$. After recording a sufficient number of photons, Bob publicly announces the sequence of analyzers he has used, but not his results. Alice compares her sequence of polarizers to Bob's analyzers and publicly gives him the list of polarizers compatible with his analyzers. The bits corresponding to incompatible analyzers and polarizers are rejected $(-)$, and then Alice and Bob are certain that the values of the other bits are the same. These are the bits which will be used to construct the key, and they are known only to Bob and Alice, because an outsider knows only the list of orientations and not the results! The protocol we have described is called BB84, from the names of its inventors Bennett and Brassard.

We still need to be sure that the message has not been intercepted and that the key it contains can be used without risk. Alice and Bob choose at random a subset of their key and compare their choices publicly. The consequence of interception of the photons by Eve will be a reduction of the correlation between the values of their bits. Let us suppose, for example, that Alice sends a photon

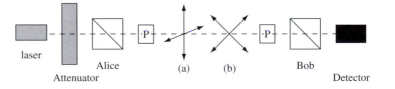

Figure 2.6 Schematical depiction of the BB84 protocol. A laser beam is attenuated such that it sends individual photons. A birefringent plate selects the polarization, which can be rotated by means of Pockels cells P. The photons are either vertically/horizontally polarized (a) or polarized at $\pm 45°$ (b).

Alice's polarizers	↑	↔	╱	↑	╱	╱	╲	↑	╲
Sequence of bits	0	1	1	0	1	1	0	0	0
Bob's polarizers	⊹	✕	⊹	⊹	✕	✕	⊹	⊹	✕
Bob's measurements	0	0	1	0	1	1	0	0	0
Retained bits	0	–	–	0	1	1	–	0	0

Figure 2.7 Quantum cryptography: transmission of polarized photons between Bob and Alice.

polarized along Ox. If Eve intercepts it using a polarizer oriented along Ox', and if the photon passes through her analyzer, she does not know that this photon was initially polarized along Ox. She then resends to Bob a photon polarized in the direction Ox', and in 50% of cases Bob will not obtain the correct result. Since Eve has one chance in two of orienting her analyzer in the right direction, Alice and Bob will record a difference in 25% of the cases and will conclude that the message has been intercepted. It is the use of two complementary bases which allows the maximal security (Exercise 2.6.4). To summarize: the security of the protocol depends on the fact that Eve cannot find out the polarization state of a photon unless she knows beforehand the basis in which it was prepared.

Of course, this discussion has been simplified considerably. It does not take into account the possibility of errors which must be corrected thanks to a classical error-correcting code, while another classical process, called privacy amplification, ensures the secrecy of the key. Moreover, the scheme should be realized using single photons and not packets of coherent states like those produced by the attenuated laser of Fig. 2.6, which are less secure, but are used for practical reasons. [11] The *quantum bit error rate* (QBER) is simply the probability that Bob measures the wrong value of the polarization when he knows Alice's basis, for example the probability that he measures a y polarization while a photon polarized along Ox was sent by Alice. It can be shown that the QBER must be less than 11% if Alice and Bob want to obtain a reliable key. The QEBR q must obey

$$q \le q_0 \quad q_0 \log q_0 + (1 - q_0) \log(1 - q_0) = \frac{1}{2},$$

---

[11] For example, an attenuated laser pulse contains typically 0.1 photon on the average. It can then be shown that a nonempty pulse has a 5% probability of containing two photons, a fact which can be exploited by Eve. In the case of transmission of isolated photons, the quantum no-cloning theorem (Section 4.3) guarantees that it is impossible for Eve to trick Bob, even if the error rate can be decreased to less than 25% by using a more sophisticated interception technique.

where log is a base 2 logarithm. If Eve is limited on attacking individual qubits, [12] the QBER must be less than 15%:

$$q \le q'_0 = \frac{1 - 1/\sqrt{2}}{2}.$$

A prototype has recently been realized for transmissions of several kilometers through air. When an optical fiber is used it is difficult to control the direction of the polarization over large distances, and so in that case a different physical support is needed to implement the BB84 protocol. Transmission over about a hundred kilometers has been achieved using optical fibers, and two versions of the device are available on the market.

2.6 Exercises

2.6.1 Determination of the polarization of a light wave

1. The polarization of a light wave is described by two complex parameters

$$\lambda = \cos\theta\, e^{i\delta_x}, \qquad \mu = \sin\theta\, e^{i\delta_y}$$

satisfying $|\lambda|^2 + |\mu|^2 = 1$. More explicitly, the electric field is

$$E_x(t) = E_0 \cos\theta \cos(\omega t - \delta_x) = E_0 \operatorname{Re}\left(\cos\theta\, e^{i\delta_x}\, e^{-i\omega t}\right),$$

$$E_y(t) = E_0 \sin\theta \cos(\omega t - \delta_y) = E_0 \operatorname{Re}\left(\sin\theta\, e^{i\delta_y}\, e^{-i\omega t}\right).$$

Determine the axes of the ellipse traced by the tip of the electric field vector and the direction in which it is traced.

2. This light wave is made to pass through a polarizing filter whose axis is parallel to Ox. Show that measurement of the intensity at the exit of the filter allows θ to be determined.

3. Now the filter is oriented such that its axis makes an angle of $\pi/4$ with Ox. What is the reduction of the intensity at the exit from the filter? Show that this second measurement permits determination of the phase difference $\delta = \delta_y - \delta_x$.

2.6.2 The (λ, μ) polarizer

1. In (2.7) we use complex notation:

$$E_x(t) = E_{0x} \cos(\omega t - \delta_x) = \operatorname{Re}\left(E_{0x}\, e^{i\delta_x}\, e^{-i\omega t}\right) = \operatorname{Re}\left(\mathcal{E}_x\, e^{-i\omega t}\right),$$

$$E_y(t) = E_{0y} \cos(\omega t - \delta_y) = \operatorname{Re}\left(E_{0y}\, e^{i\delta_y}\, e^{-i\omega t}\right) = \operatorname{Re}\left(\mathcal{E}_y\, e^{-i\omega t}\right).$$

---

[12] That is, she is not allowed to store many qubits, which would permit coherent attacks on many qubits.

Let the two numbers λ real and μ complex be parametrized as

$$\lambda = \cos\theta, \quad \mu = \sin\theta\,e^{i\eta}.$$

A (λ, μ) polarizer is constructed of three elements.

- A first birefringent plate which changes the phase of $\mathcal{E}_y$ by $-\eta$, leaving $\mathcal{E}_x$ unchanged:

$$\mathcal{E}_x \to \mathcal{E}_x^{(1)} = \mathcal{E}_x, \quad \mathcal{E}_y \to \mathcal{E}_y^{(1)} = \mathcal{E}_y\,e^{-i\eta}.$$

- A linear polarizer which projects onto the unit vector $\hat{n}_\theta = (\cos\theta, \sin\theta)$:

$$\vec{\mathcal{E}}^{(1)} \to \vec{\mathcal{E}}^{(2)} = \left(\mathcal{E}_x^{(1)}\cos\theta + \mathcal{E}_y^{(1)}\sin\theta\right)\hat{n}_\theta$$

$$= \left(\mathcal{E}_x\cos\theta + \mathcal{E}_y\sin\theta\,e^{-i\eta}\right)\hat{n}_\theta.$$

- A second birefringent plate which leaves $\mathcal{E}_x^{(2)}$ unchanged and changes the phase of $\mathcal{E}_y^{(2)}$ by η:

$$\mathcal{E}_x^{(2)} \to \mathcal{E}_x' = \mathcal{E}_x^{(2)}, \quad \mathcal{E}_y^{(2)} \to \mathcal{E}_y' = \mathcal{E}_y^{(2)}\,e^{i\eta}.$$

The combination of all three operations is represented as $\vec{\mathcal{E}} \to \vec{\mathcal{E}}'$. Calculate the components $\mathcal{E}_x'$ and $\mathcal{E}_y'$ as functions of $\mathcal{E}_x$ and $\mathcal{E}_y$.

2. Vectors of $\mathcal{H}$ which are not normalized, $|\mathcal{E}\rangle$ and $|\mathcal{E}'\rangle$, are defined as

$$|\mathcal{E}\rangle = \mathcal{E}_x|x\rangle + \mathcal{E}_y|y\rangle, \quad |\mathcal{E}'\rangle = \mathcal{E}_x'|x\rangle + \mathcal{E}_y'|y\rangle.$$

Show that the operation $|\mathcal{E}\rangle \to |\mathcal{E}'\rangle$ is a projection:

$$|\mathcal{E}'\rangle = \mathcal{P}_\Phi|\mathcal{E}\rangle,$$

where $\mathcal{P}_\Phi$ is the projector onto the vector

$$|\Phi\rangle = \lambda|x\rangle + \mu|y\rangle.$$

3. Show that a photon with state vector $|\Phi\rangle$ is transmitted by the (λ, μ) polarizer with unit probability, and that a photon of state vector

$$|\Phi_\perp\rangle = -\mu^*|x\rangle + \lambda^*|y\rangle$$

is stopped by this polarizer.

2.6.3 Circular polarization and the rotation operator

1. Justify the following expressions for the states $|R\rangle$ and $|L\rangle$ respectively representing right- and left-handed polarized photons:

$$|R\rangle = \frac{1}{\sqrt{2}}(|x\rangle + i|y\rangle), \quad |L\rangle = \frac{1}{\sqrt{2}}(|x\rangle - i|y\rangle),$$

where $|x\rangle$ and $|y\rangle$ are the state vectors of photons linearly polarized along Ox and Oy. Hint: what is the electric field of a circularly polarized light wave? Write down the matrix form of the projectors $\mathcal{P}_R$ and $\mathcal{P}_L$ onto the states $|R\rangle$ and $|L\rangle$ in the basis $\{|x\rangle, |y\rangle\}$.

2. We define the states $|\theta\rangle$ and $|\theta_\perp\rangle$ (2.19) representing photons linearly polarized along directions making an angle θ with Ox and Oy, respectively, and also the states

$$|R'\rangle = \frac{1}{\sqrt{2}}(|\theta\rangle + i|\theta_\perp\rangle), \quad |L'\rangle = \frac{1}{\sqrt{2}}(|\theta\rangle - i|\theta_\perp\rangle).$$

How are $|R'\rangle$ and $L'\rangle$ related to $|R\rangle$ and $|L\rangle$? Do these state vectors represent physical states different from $|R\rangle$ and $|L\rangle$? If not, then why not?

3. We construct the Hermitian operator

$$\Sigma = \mathcal{P}_R - \mathcal{P}_L.$$

What is the action of Σ on the vectors $|R\rangle$ and $|L\rangle$? Determine the action of $\exp(-i\theta\Sigma)$ on these vectors.

4. Write the matrix representing Σ in the basis $\{|x\rangle, |y\rangle\}$. Show that $\Sigma^2 = I$ and recover $\exp(-i\theta\Sigma)$. By comparing with question **2**, give the physical interpretation of the operator $\exp(-i\theta\Sigma)$.

2.6.4 An optimal strategy for Eve?

1. Let us suppose that Eve analyzes the polarization of a photon sent by Alice using an analyzer oriented as $\updownarrow$. If Alice orients her polarizer as $\updownarrow$, the probability that Eve measures the value of the qubit as $+1$ is 100% when Alice sends a qubit $+1$, but only 50% when Alice uses a $\diagdown$ polarizer. The probability that Eve measures $+1$ when Alice sends $+1$ then is

$$p = \frac{1}{2} + \frac{1}{2}\left(\frac{1}{2}\right) = \frac{3}{4}.$$

Let us suppose that Eve orients her analyzer in a direction making an angle ϕ with Ox. Show that the probability $p(\phi)$ for Eve to measure $+1$ when Alice sends $+1$ is now

$$p(\phi) = \frac{1}{4}(2 + \cos 2\phi + \sin 2\phi).$$

Show that for the optimal choice $\phi = \phi_0 = \pi/8$

$$p(\phi_0) \simeq 0.854,$$

a larger value than before. Would it have been possible to predict without calculation that the optimal value must be $\phi = \phi_0 = \pi/8$? However, as explained in Section 7.2, the information gain of Eve is less than with the naive strategy.

2. Suppose that instead of using a basis $|\pm\pi/4\rangle$, Alice uses a $\{|\theta\rangle, |\theta_\perp\rangle\}$ basis. Show that the probability that Eve makes a wrong guess is now

$$\mathsf{p} = \frac{1}{4}\sin^2(2\theta).$$

Thus, the use of complementary bases maximizes Eve's error rate.

2.6.5 Heisenberg inequalities

1. Let us take two Hermitian operators A and B. Show that their *commutator* $[A, B]$ is anti-Hermitian,

$$[A, B] := AB - BA = iC,$$

where C is Hermitian: $C = C^\dagger$.

2. The expectation values of A and B are defined as

$$\langle A \rangle_\varphi = \langle \varphi | A\varphi \rangle, \quad \langle B \rangle_\varphi = \langle \varphi | B\varphi \rangle,$$

and the *dispersions* $\Delta_\varphi A$ and $\Delta_\varphi B$ *in the state* $|\varphi\rangle$ as

$$(\Delta_\varphi A)^2 = \langle A^2 \rangle_\varphi - (\langle A \rangle_\varphi)^2 = \langle (A - \langle A \rangle_\varphi I)^2 \rangle_\varphi,$$
$$(\Delta_\varphi B)^2 = \langle B^2 \rangle_\varphi - (\langle B \rangle_\varphi)^2 = \langle (B - \langle B \rangle_\varphi I)^2 \rangle_\varphi.$$

Finally, we define Hermitian operators of zero expectation value (which are *a priori specific to the state* $|\varphi\rangle$) as

$$A_0 = A - \langle A \rangle_\varphi I, \quad B_0 = B - \langle B \rangle_\varphi I.$$

What is their commutator? The norm of the vector

$$(A_0 + i\lambda B_0)|\varphi\rangle,$$

where λ is chosen to be real, must be positive:

$$\|(A_0 + i\lambda B_0)|\varphi\rangle\| \geq 0.$$

Derive the Heisenberg inequality

$$(\Delta_\varphi A)(\Delta_\varphi B) \geq \frac{1}{2}\left|\langle C \rangle_\varphi\right|.$$

Care must be taken in interpreting this inequality. It implies that if a large number of quantum systems are prepared in the state $|\varphi\rangle$, and if their expectation values and dispersions $\{\langle A \rangle_\varphi, \Delta_\varphi A\}$, $\{\langle B \rangle_\varphi, \Delta_\varphi B\}$, and $\langle C \rangle_\varphi$ are measured in *independent* experiments, then these expectation values will obey the Heisenberg inequality. In contrast to what is sometimes found in the literature, the dispersions $\Delta_\varphi A$ and $\Delta_\varphi B$ are not at all associated with the experimental errors. There is nothing which *a priori* prevents $\langle A \rangle_\varphi$, for example, from being measured with an accuracy better than $\Delta_\varphi A$.

3. The position and momentum operators X and P (in one dimension) obey the commutation relation

$$[X, P] = i\hbar I,$$

where $\hbar$ is the Planck constant, $\hbar = 1.054 \times 10^{-34} \, \text{J s}$. Show that this commutation relation cannot be satisfied by operators acting in a Hilbert space of finite dimension. Hint: study the trace of this equation. From question **2** derive the Heisenberg inequality

$$\Delta X \Delta P \geq \frac{1}{2}\hbar.$$

2.7 Further reading

For additional information about light polarization and photons, the reader can consult Le Bellac (2006), Chapter 3, Lévy-Leblond and Balibar (1990), Chapter 4, and Hey and Walters (2003), Chapter 8. For the general principles of quantum mechanics see, for example, Nielsen and Chuang (2000), Chapter 2, which contains an elegant proof of the spectral decomposition theorem (Section 2.4). Some examples of determining a trajectory without perturbation in the Young slit experiment are given by Englert *et al.* (1991) and Dürr *et al.* (1998). A recent review on quantum cryptography containing numerous references to earlier work is that of Gisin *et al.* (2002). Popular accounts of quantum cryptography can be found in Bennett *et al.* (1992) and in Johnson (2003), Chapter 9. A very readable book on cryptography is Singh (2000).

3

Manipulating qubits

In the preceding chapter we studied a qubit at a particular instant of time and we neglected, for example, the qubit evolution between preparation and measurement. As we have seen, in a Hilbert space $\mathcal{H}$ with an orthonormal basis formed of two vectors $|0\rangle$ and $|1\rangle$, this qubit is described by a normalized vector $|\varphi\rangle$:

$$|\varphi\rangle = \lambda|0\rangle + \mu|1\rangle, \qquad |\lambda|^2 + |\mu|^2 = 1. \tag{3.1}$$

In this chapter, we want to examine the time evolution of this qubit, in order to understand how it can be manipulated. We shall see that Rabi oscillations (Section 3.3) provide the basic mechanism that allows us to manipulate qubits.

3.1 The Bloch sphere, spin 1/2

Before turning to time evolution, let us give a somewhat more general description of a qubit and of its physical realizations. In writing down (3.1) we have used an orthonormal basis $\{|0\rangle, |1\rangle\}$ of $\mathcal{H}$, and the coefficients λ and μ can be parametrized, taking into account the arbitrariness of the phase, as

$$\lambda = e^{-i\phi/2} \cos\frac{\theta}{2}, \qquad \mu = e^{i\phi/2} \sin\frac{\theta}{2}. \tag{3.2}$$

The two angles θ and ϕ can be taken as the polar and azimuthal angles which parametrize the location of a point on the surface of a sphere of unit radius called the *Bloch sphere* (or the Poincaré sphere for the photon); see Fig. 3.1.

Returning to the photon polarization and identifying $|0\rangle \to |x\rangle$ and $|1\rangle \to |y\rangle$, the states $|x\rangle$ and $|y\rangle$ correspond to the north and south poles of the sphere:

$$|x\rangle: \ \theta = 0, \ \phi \text{ undetermined}, \qquad |y\rangle: \ \theta = \pi, \ \phi \text{ undetermined}.$$

Circular polarizations correspond to points on the equator:

$$|R\rangle: \ \theta = \frac{\pi}{2}, \ \phi = \frac{\pi}{2}, \qquad |L\rangle: \ \theta = \frac{\pi}{2}, \ \phi = -\frac{\pi}{2}.$$

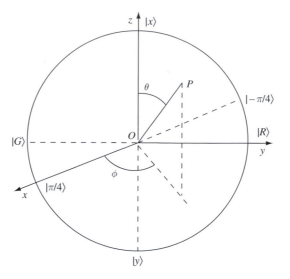

Figure 3.1 The Bloch sphere. The points on the Bloch sphere corresponding to the photon polarization bases $\{|x\rangle, |y\rangle)\}$, $\{|R\rangle, |L\rangle\}$, and $\{|\theta = \pi/4\rangle,$ $|\theta = -\pi/4\rangle\}$ are shown.

Another important physical realization of the qubit is spin 1/2. Let us introduce the subject by discussing a very well known phenomenon. A small magnetized needle is an approximate realization of what physicists call a *magnetic dipole*, characterized by a *magnetic dipole moment*, or simply *magnetic moment*, $\vec{\mu}$, which is an (axial) vector of $\mathbb{R}^3$. When placed in a magnetic field $\vec{B}$, this needle aligns itself in the direction of the field, just as the needle of a compass aligns itself with the Earth's magnetic field. The reason for this alignment is the following. The energy E of a magnetic dipole in a field $\vec{B}$ is

$$E = -\vec{\mu} \cdot \vec{B}, \tag{3.3}$$

and the minimum energy [1] is obtained when $\vec{\mu}$ is parallel to and in the same direction as $\vec{B}$. When the field is not uniform, the dipole moves toward the region where the field has the largest absolute value so as to minimize its energy. In summary, a dipole is subject to a torque which tends to align it with the field, and to a force which tends to make it move under the influence of a field *gradient*.

NMR (Nuclear Magnetic Resonance) and its derivative MRI (Imaging by (Nuclear) Magnetic Resonance [2]) are based on the fact that the proton [3] possesses a magnetic moment which can take two *and only two* values along the direction

---

[1] A physical system always struggles to reach a state of minimum energy (more correctly, minimum free energy).

[2] The adjective "nuclear" has been suppressed in order not to frighten the public...

[3] In fact, other nuclei of spin 1/2 such as ^{13}C, ^{19}F, and so on are also used in NMR; see Sec. 6.2. Only protons are used in MRI.

of a magnetic field. In other words, the component $\vec{\mu} \cdot \hat{n}$ of $\vec{\mu}$ along any axis $\hat{n}$ takes only two values, and this property characterizes a *spin 1/2 particle*. Experimentally, this can be seen as follows. A beam of protons [4] is sent into a magnetic field pointing in a direction $\hat{n}$ perpendicular to the beam direction. It is observed that the beam splits into two sub-beams, one deflected in the direction $\hat{n}$, and the other in the opposite direction $-\hat{n}$. This is the Stern–Gerlach experiment (Fig. 3.2, with $\hat{n} \parallel Oz$), which is in its principle a close analog of the separation of a ray of natural light into two rays by a birefringent crystal. The analog of a polarizer–analyzer experiment using spin 1/2 can also be imagined (Fig. 3.3). However, it should be noted that that the crossed polarizer–analyzer situation corresponds to

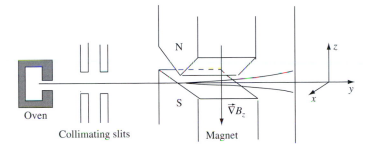

Figure 3.2 The Stern–Gerlach experiment. Silver atoms leaving an oven are collimated and pass through the gap of a magnet constructed such that the field is nonuniform with the gradient pointing in the $-z$ direction. It is actually the electron magnetic moment, which is a thousand times larger than the proton magnetic moment, which is responsible for the deflection.

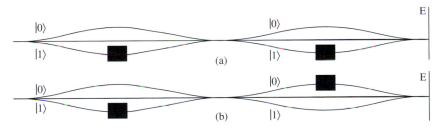

Figure 3.3 Crossed polarizers for spin 1/2. In case (a) 100% of the spins are transmitted by the second Stern–Gerlach apparatus, while 0% are transmitted in case (b). The two Stern–Gerlach filters select spin 1/2 particles in state $|0\rangle$ (upper beam) and state $|1\rangle$ (lower beam).

---

[4] This is a thought experiment. It is actually necessary to use neutral atoms rather than protons, as in Fig. 3.2; otherwise, the effects will be masked by forces due to the charges. Moreover, nuclear magnetism is too weak to be seen in such an experiment.

$\theta = \pi$ rather than $\theta = \pi/2$ as in the case of photons.[5] We construct a basis of $\mathcal{H}$ taking as the basis vectors $|0\rangle$ and $|1\rangle$, which correspond to states prepared by a magnetic field parallel to Oz. According to (3.1) and (3.2), the most general spin 1/2 state is

$$|\varphi\rangle = e^{-i\phi/2} \cos \frac{\theta}{2} |0\rangle + e^{i\phi/2} \sin \frac{\theta}{2} |1\rangle, \tag{3.4}$$

and it can be shown[6] that this state is the one selected by a magnetic field parallel to $\hat{n}$ with

$$\hat{n} = (\sin \theta \cos \phi, \sin \theta \sin \phi, \cos \theta). \tag{3.5}$$

In the spin 1/2 case, the Bloch sphere has an obvious geometrical interpretation: the spin 1/2 described by the vector (3.4) points in the direction $\hat{n}$.

We have seen that the physical properties of qubits are represented by Hermitian operators acting in a two-dimensional space. A convenient basis for these operators is that of the *Pauli matrices*:

$$\sigma_1 (\text{or } \sigma_x) = \begin{pmatrix} 0 & 1 \\ 1 & 0 \end{pmatrix}, \quad \sigma_2 (\text{or } \sigma_y) = \begin{pmatrix} 0 & -i \\ i & 0 \end{pmatrix}, \quad \sigma_3 (\text{or } \sigma_z) = \begin{pmatrix} 1 & 0 \\ 0 & -1 \end{pmatrix} \tag{3.6}$$

These matrices are Hermitian (and also unitary), and any 2×2 Hermitian matrix M can be written as

$$M = \lambda_0 I + \sum_{i=1}^{3} \lambda_i \sigma_i \tag{3.7}$$

with real coefficients. The Pauli matrices possess the following important properties:

$$\sigma_i^2 = I, \quad \sigma_1 \sigma_2 = i\sigma_3, \quad \sigma_2 \sigma_3 = i\sigma_1, \quad \sigma_3 \sigma_1 = i\sigma_2. \tag{3.8}$$

The states $|0\rangle$ and $|1\rangle$ are eigenvectors of σ_z with the eigenvalues ± 1:

$$|0\rangle = \begin{pmatrix} 1 \\ 0 \end{pmatrix}, \quad |1\rangle = \begin{pmatrix} 0 \\ 1 \end{pmatrix}, \tag{3.9}$$

and it can be verified immediately that the vector $|\varphi\rangle$ (3.4) is an eigenvector of

$$\vec{\sigma} \cdot \hat{n} = \sigma_x n_x + \sigma_y n_y + \sigma_z n_z = \begin{pmatrix} \cos \theta & e^{-i\phi} \sin \theta \\ e^{i\phi} \sin \theta & -\cos \theta \end{pmatrix} \tag{3.10}$$

---

[5] The photon has spin 1, and not 1/2! The rotation operator of a photon (Exercise 2.6.3) can be compared with that of a spin 1/2 (see Exercise 3.5.1), and it will be seen that it is the angle θ which arises in the first case and the angle $\theta/2$ in the second. A note for physicists: a massive particle of spin 1 possesses three polarization states, not two. An analysis performed by Wigner in 1939 shows that a zero-mass particle like the photon has only *two* polarization states no matter what its spin is.

[6] This is a consequence of the invariance under rotation; see Exercise 3.5.1.

with eigenvalue $+1$. We also note that the vector $\langle \vec{\sigma} \rangle$, the expectation value of the spin in the state (3.4), is given by

$$\langle \vec{\sigma} \rangle = (\langle \sigma_x \rangle, \langle \sigma_y \rangle, \langle \sigma_x \rangle) \tag{3.11}$$

and points along $\hat{n}$.

We have just demonstrated the physical realization of a qubit by a spin 1/2, but there exist many other realizations, such as by a two-level atom, see the following section. In any case, the Hilbert space always has dimension 2, and the state of a qubit can always be represented by a point on the Bloch sphere.

3.2 Dynamical evolution

Now let us explicitly introduce the time, assuming that (3.1) holds at $t = 0$:

$$|\varphi(t=0)\rangle = \lambda(t=0)|0\rangle + \mu(t=0)|1\rangle, \qquad \lambda(t=0) = \lambda, \ \mu(t=0) = \mu. \tag{3.12}$$

Principle 3 We shall assume that the transformation

$$|\varphi(0)\rangle \to |\varphi(t)\rangle$$

is linear and that the norm of $|\varphi\rangle$ remains equal to one [7]

$$|\varphi(t)\rangle = \lambda(t)|0\rangle + \mu(t)|1\rangle, \tag{3.13}$$

$$|\lambda(t)|^2 + |\mu(t)|^2 = 1. \tag{3.14}$$

The transformation $|\varphi(0)\rangle \to |\varphi(t)\rangle$ is then a *unitary transformation* $U(t, 0)$ (a *unitary operator* U satisfies $U^{-1} = U^\dagger$; in a finite dimensional space, a linear operator which preserves the norm, called an isometry, is also a unitary operator):

$$|\varphi(t)\rangle = U(t, 0)|\varphi(t=0)\rangle.$$

In general,

$$|\varphi(t_2)\rangle = U(t_2, t_1)|\varphi(t_1)\rangle, \qquad U^\dagger(t_2, t_1) = U^{-1}(t_2, t_1). \tag{3.15}$$

Moreover, U must satisfy the group property:

$$U(t_2, t_1) = U(t_2, t')U(t', t_1), \tag{3.16}$$

---

[7] This second condition seems to be a natural consequence of state vector normalization, but in fact it involves the assumption that *all* the quantum degrees of freedom are included in $\mathcal{H}$: the evolution is not in general unitary when the qubit is only a part of a larger quantum system for which the Hilbert space of states is larger than $\mathcal{H}$, see Section 4.4.

and, finally, $U(t, t) = I$. We use the group property and Taylor expansion for infinitesimal dt to write

$$U(t + dt, t_0) = U(t + dt, t)U(t, t_0),$$

$$U(t + dt, t_0) \simeq U(t, t_0) + dt \frac{d}{dt} U(t, t_0),$$

$$U(t + dt, t)U(t, t_0) \simeq \left[I - \frac{i}{\hbar} dt \hat{H}(t) \right] U(t, t_0),$$

where we have defined the operator $\hat{H}(t)$, the *Hamiltonian*, as

$$\hat{H}(t) = i\hbar \left. \frac{dU(t', t)}{dt'} \right|_{t' = t}. \tag{3.17}$$

The constant $\hbar = 1.05 \times 10^{-34}$ J s was first introduced by Planck and is called Planck's constant. It relates energy E and frequency ω according to the *Planck–Einstein formula* $E = \hbar\omega$. The presence of the factor i ensures that $\hat{H}(t)$ is a Hermitian operator, while the presence of $\hbar$ implies that $\hat{H}$ has the dimension of an energy. In fact,

$$I = U^\dagger(t + dt, t)U(t + dt, t) \simeq \left[I + \frac{i}{\hbar} dt \hat{H}^\dagger(t) \right] \left[I - \frac{i}{\hbar} dt \hat{H}(t) \right]$$

$$\simeq I + \frac{i}{\hbar} dt (\hat{H}^\dagger - \hat{H}),$$

which implies that $\hat{H} = \hat{H}^\dagger$. From the above we see that the *evolution equation* (also called the *Schrödinger equation*) is

$$\boxed{i\hbar \frac{dU(t, t_0)}{dt} = \hat{H}(t)U(t, t_0)} \tag{3.18}$$

Since $\hat{H}$ is a Hermitian operator, this is a physical property, and in fact $\hat{H}$ is just the *energy operator* of the system. In the often encountered case where the physics is invariant under time translation, the operator $U(t_2, t_1)$ depends only on the *difference* $(t_2 - t_1)$ and $\hat{H}$ is independent of time.

Let us illustrate this for the example of NMR (or MRI). In the first stage the spins 1/2 are placed in a strong, time independent magnetic field $\vec{B}_0$ (B_0 is a few teslas and 1 tesla $= 10^4$ gauss, about 10^4 times as strong as the Earth's magnetic field, which is why it is better not to wear one's watch when undergoing an MRI scan!). The Hamiltonian is then time independent, and since it is Hermitian it can be diagonalized in a certain basis:

$$\hat{H} = \begin{pmatrix} \hbar\omega_A & 0 \\ 0 & \hbar\omega_B \end{pmatrix}, \tag{3.19}$$

where $\hbar\omega_A$ and $\hbar\omega_B$ are the *energy levels* of the spin 1/2. If the magnetic field is parallel to Oz, the eigenvectors of $\hat{H}$ are just the basis vectors $|0\rangle$ and $|1\rangle$; see Box 3.1. Since $\hat{H}$ is independent of time, the evolution equation (3.18),

$$i\hbar \frac{dU}{dt} = \hat{H}U,$$

can be integrated directly to give

$$U(t, t_0) = \exp[-i\hat{H}(t - t_0)/\hbar], \qquad (3.20)$$

or, in the basis in which $\hat{H}$ is diagonal,

$$U(t, t_0) = \begin{pmatrix} e^{-i\omega_A(t-t_0)} & 0 \\ 0 & e^{-i\omega_B(t-t_0)} \end{pmatrix}. \qquad (3.21)$$

If $|\varphi(t=0)\rangle$ is given by

$$|\varphi(t=0)\rangle = \lambda|0\rangle + \mu|1\rangle,$$

then the state vector $|\varphi(t)\rangle$ at time t is

$$|\varphi(t)\rangle = e^{-i\omega_A t}\lambda|0\rangle + e^{-i\omega_B t}\mu|1\rangle \qquad (3.22)$$

or

$$\lambda(t) = e^{-i\omega_A t}\lambda, \qquad \mu(t) = e^{-i\omega_B t}\mu.$$

The time evolution is *deterministic* and it keeps the memory of the initial conditions λ and μ. Owing to the arbitrariness of the phase, the only quantity which actually is physically relevant to the evolution is the difference

$$\omega_0 = \omega_B - \omega_A, \qquad (3.23)$$

so that it is also possible to write $\hat{H}$ as

$$\hat{H} = -\frac{1}{2}\begin{pmatrix} \hbar\omega_0 & 0 \\ 0 & -\hbar\omega_0 \end{pmatrix}.$$

The quantity ω_0 plays an important role and is called the *resonance frequency*, and $\hbar\omega_0$ the *resonance energy*. By solving the equations of motion of a classical spin, it can be shown that the classical spin precesses about $\vec{B}_0$ with an angular frequency ω_0, the *Larmor frequency*.

Let us take this opportunity to mention another physical realization of a qubit, namely, a *two-level atom*. An atom possesses a large number of energy levels, but if we are interested in the effect of laser light on this atom, it is often possible to restrict ourselves to two particular levels, in general, the ground state ω_A and an excited state ω_B, $\omega_B > \omega_A$. This is referred to as the model of the two-level atom and it is very widely used in atomic physics. If the atom is raised to its excited state, it returns spontaneously to its ground state by emitting a photon of frequency $\omega_0 = \omega_B - \omega_A$. If the atom in its ground state is hit with a laser beam of frequency $\omega \simeq \omega_0$, a *resonance phenomenon* is observed: the laser light will be absorbed more strongly the closer ω is to ω_0, a phenomenon which is analogous to that described in the following section in the case of spin 1/2.

3.3 Manipulating qubits: Rabi oscillations

Box 3.1: Interaction of a spin 1/2 with a magnetic field

An elementary calculation of classical physics shows that the magnetic moment $\vec{\mu}$ of a rotating charged system is proportional to its angular momentum $\vec{J}$, $\vec{\mu} = (\gamma/\hbar)\vec{J}$, where γ is called the *gyromagnetic ratio*. The proton spin is in fact an intrinsic angular momentum, rather as though the proton were spinning on its axis like a top. However, this classical image of the proton spin should be used with care, as it can be completely incorrect in the interpretation of certain phenomena; only a quantum description actually permits a real understanding of spin. Intrinsic angular momentum is a vectorial physical property with which there must be an associated Hermitian operator (more precisely, three Hermitian operators, one for each component). The proton spin is associated with the operator $\hbar\vec{\sigma}/2$. Note that the dimensionality is correct, because an angular momentum has the same dimension as $\hbar$. The magnetic moment, also a vector, is associated with a corresponding operator which must be proportional to the intrinsic angular momentum, because the only vector (actually, axial vector) at our disposal is $\vec{\sigma}$:

$$\vec{\mu} = \frac{1}{2}\gamma_p\vec{\sigma}, \qquad \gamma_p = 5.59\frac{q_p\hbar}{2m_p},$$

where γ_p is the gyromagnetic ratio of the proton, q_p is the proton charge, and m_p is the proton mass. The numerical value of γ_p must be taken from experiment,[8] and at present there is no reliable way to calculate it theoretically.[9]

As we shall see in Chapter 5, in quantum computing it is necessary to be able to transform a state, $|0\rangle$ for example, of a qubit into a linear superposition of $|0\rangle$ and $|1\rangle$. Taking spin 1/2 as an example, this can be done, as we shall see, by applying to the spin a constant magnetic field $\vec{B}_0$ parallel to Oz and a magnetic field $\vec{B}_1(t)$ rotating in the xOy plane with angular velocity ω:

$$\vec{B}_1(t) = B_1(\hat{x}\cos\omega t - \hat{y}\sin\omega t).$$

---

[8] The magnetic moment $\mu = 1.4 \times 10^{-28}$ J/T.

[9] In principle, it should be possible to calculate γ_p using the theory of strong interactions, QCD (Quantum ChromoDynamics). In practice, this calculation has to be done numerically (using lattice QCD), and the present accuracy is very far from permitting a good estimate of γ_p.

The Hamiltonian of the proton magnetic moment in a magnetic field is written by analogy with (3.3), since $\hat{H}$ is the energy operator:

$$\hat{H} = -\vec{\mu} \cdot \vec{B} = -\frac{1}{2} \gamma_p \vec{\sigma} \cdot \vec{B}.$$

The magnetic field used in NMR is

$$\vec{B} = B_0 \hat{z} + B_1 (\hat{x} \cos \omega t - \hat{y} \sin \omega t).$$

We define $\hbar \omega_0 = \gamma_p B_0$ and $\hbar \omega_1 = \gamma_p B_1$, and then the Hamiltonian becomes

$$\hat{H}(t) = -\frac{1}{2} \gamma_p B_0 \sigma_z - \frac{1}{2} \gamma_p B_1 (\sigma_x \cos \omega t - \sigma_y \sin \omega t)$$

$$= -\frac{\hbar}{2} \omega_0 \sigma_z - \frac{\hbar}{2} \omega_1 (\sigma_x \cos \omega t - \sigma_y \sin \omega t),$$

and we find (3.24) using the explicit form (3.6) of the Pauli matrices. If $\vec{B}_1 = 0$, the Hamiltonian is time independent and its eigenvectors are $|0\rangle$ and $|1\rangle$, with eigenvalues $-\hbar \omega_0 / 2$ and $+\hbar \omega_0 / 2$, respectively.

Let a spin 1/2 be placed in a classical magnetic field with a periodic component as in Box 3.1:

$$\vec{B} = \vec{B}_0 \hat{z} + B_1 (\hat{x} \cos \omega t - \hat{y} \sin \omega t).$$

The form of $\hat{H}(t)$ then is (see Box 3.1 for the justification of (3.24))

$$\hat{H}(t) = -\frac{\hbar}{2} \begin{pmatrix} \omega_0 & \omega_1 e^{i\omega t} \\ \omega_1 e^{-i\omega t} & -\omega_0 \end{pmatrix}, \qquad (3.24)$$

where ω_1 is proportional to B_1 and can therefore be adjusted at will. The frequency ω_1 is called the *Rabi frequency*. The evolution equation (3.18) still needs to be solved. It is easily transformed into a system of two coupled first-order differential equations for $\lambda(t)$ and $\mu(t)$, which can be solved without difficulty (see Box 3.2 and Exercise 3.5.2). The result can be expressed as follows. If at time $t = 0$ the qubit is in the state $|0\rangle$, at time t it will have a probability $\mathsf{p}_{0 \to 1}(t)$ of being found in the state $|1\rangle$ given by

$$\mathsf{p}_{0 \to 1}(t) = \left(\frac{\omega_1}{\Omega} \right)^2 \sin^2 \frac{\Omega t}{2}, \qquad \Omega = \sqrt{(\omega - \omega_0)^2 + \omega_1^2}. \qquad (3.25)$$

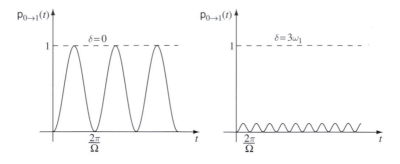

Figure 3.4 Rabi oscillations. The *detuning* δ is defined as $\delta = \omega - \omega_0$.

This is the phenomenon of *Rabi oscillations* (Fig. 3.4). The oscillation between the levels $|0\rangle$ and $|1\rangle$ has maximum amplitude for $\omega = \omega_0$, that is, at *resonance*:

$$p_{0\to 1}(t) = \sin^2 \frac{\omega_1 t}{2}, \qquad \omega = \omega_0. \tag{3.26}$$

To go from the state $|0\rangle$ to the state $|1\rangle$ it is sufficient to adjust the time t during which the rotating field acts:

$$\frac{\omega_1 t}{2} = \frac{\pi}{2}, \qquad t = \frac{\pi}{\omega_1}.$$

This is called a π *pulse*. If a time intermediate between 0 and π/ω_1 is chosen, we obtain a superposition of $|0\rangle$ and $|1\rangle$. In particular, if $t = \pi/2\omega_1$ we have a $\pi/2$ *pulse*:

$$|0\rangle \to \frac{1}{\sqrt{2}}(|0\rangle + |1\rangle). \tag{3.27}$$

This operation will be of crucial importance in quantum computing. The equations are essentially identical in the case of a two-level atom in the field of a laser when the generally well satisfied "rotating-wave approximation" is made. Then $\hbar\omega_0$ is the energy difference between the two atomic levels, ω is the frequency of the laser wave, and the Rabi frequency ω_1 is proportional to the product of the (transition) electric dipole moment of the atom $\vec{d}$ and the electric field $\vec{E}$ of the laser wave, $\omega_1 \propto \vec{d} \cdot \vec{E}\hbar$.

In summary, Rabi oscillations are the basic process used to manipulate qubits. These oscillations are obtained by exposing qubits to periodic electric or magnetic fields during suitably adjusted time intervals.

Box 3.2: Solution of the NMR evolution equation at resonance

Equation (3.24) can immediately be transformed into an equation for $|\varphi(t)\rangle = U(t)|\varphi(t=0)\rangle$:

$$i\hbar \frac{d|\varphi(t)\rangle}{dt} = \hat{H}(t)|\varphi(t)\rangle,$$

from which we find that $\lambda(t)$ and $\mu(t)$ obey a system of coupled differential equations:

$$i\frac{d\lambda(t)}{dt} = -\frac{\omega_0}{2}\lambda(t) - \frac{\omega_1}{2}e^{i\omega t}\mu(t),$$

$$i\frac{d\mu(t)}{dt} = -\frac{\omega_1}{2}e^{-i\omega t}\lambda(t) + \frac{\omega_0}{2}\mu(t). \tag{3.28}$$

It is convenient to define

$$\lambda(t) = \hat{\lambda}(t)e^{i\omega_0 t/2}, \qquad \mu(t) = \hat{\mu}(t)e^{-i\omega_0 t/2}. \tag{3.29}$$

The system of differential equations simplifies to become

$$i\frac{d\hat{\lambda}(t)}{dt} = -\frac{\omega_1}{2}e^{i(\omega-\omega_0)t}\hat{\mu}(t),$$

$$i\frac{d\hat{\mu}(t)}{dt} = -\frac{\omega_1}{2}e^{-i(\omega-\omega_0)t}\hat{\lambda}(t). \tag{3.30}$$

This system is easily transformed into a second-order differential equation for $\hat{\lambda}(t)$ (or $\hat{\mu}(t)$). Here we shall content ourselves with examining the case of resonance $\omega = \omega_0$ (see Exercise 3.5.2 for the general case), where

$$\frac{d^2\hat{\lambda}(t)}{dt^2} = -\frac{\omega_1^2}{4}\hat{\lambda}(t).$$

The solution of the system then is

$$\hat{\lambda}(t) = a\cos\frac{\omega_1 t}{2} + b\sin\frac{\omega_1 t}{2},$$

$$\hat{\mu}(t) = ia\sin\frac{\omega_1 t}{2} - ib\cos\frac{\omega_1 t}{2}. \tag{3.31}$$

The coefficients a and b depend on the initial conditions. Starting from, for example, the state $|0\rangle$ at time $t = 0$,

$$\lambda(t=0) = 1, \mu(t=0) = 0 \text{ or } a = 1, \ b = 0,$$

at time $t = \pi/2\omega_1$ (a $\pi/2$ pulse) we have a state which is a *linear superposition* of $|0\rangle$ and $|1\rangle$:

$$|\varphi(t)\rangle = \frac{1}{\sqrt{2}}\left(e^{i\omega_0 t/2}|0\rangle + ie^{-i\omega_0 t/2}|1\rangle\right). \tag{3.32}$$

The phase factors can be absorbed by redefining the states $|0\rangle$ and $|1\rangle$ such that (3.27) is obtained.

3.4 Principles of NMR and MRI

NMR spectroscopy is mainly used to determine the structure of complex chemical or biological molecules and for studying condensed matter in solid or liquid form. A detailed description of how NMR works would take us too far afield, and so we shall only touch upon the subject. The sample under study is placed in a uniform field $\vec{B}_0$ of several teslas, the maximum field accessible at present being about 20 T (Fig. 3.5). An NMR is characterized by the resonance frequency. [10] $\nu_0 = \omega_0/2\pi = \gamma_p B_0/(2\pi\hbar)$ for a proton: a field of 1 T corresponds to a frequency $\simeq 42.6$ MHz, and so we can speak of an NMR of 600 MHz if the field B_0 is 14 T. Owing to the Boltzmann law, the level $|0\rangle$ is more populated than the level $|1\rangle$, at least for $\gamma > 0$, which is the usual case. The ratio of the populations p_0 and p_1 at thermal equilibrium at absolute temperature T is, from the Boltzmann law,

$$\frac{p_0(t=0)}{p_1(t=0)} = \exp\left(\frac{\hbar\omega_0}{k_B T}\right), \qquad (3.33)$$

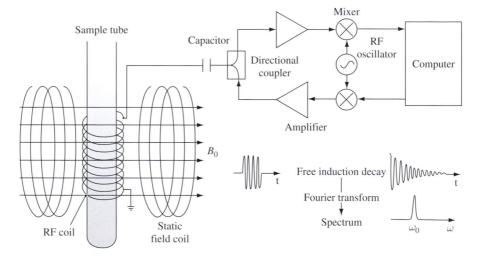

Figure 3.5 Schematics of the NMR principle. The static field $\vec{B}_0$ is horizontal and the radiofrequency field is generated by the vertical solenoid. This solenoid also serves as the signal detector (FID Free Induction Decay). The RF pulse and the signal are sketched at the lower right of the figure. The decreasing exponential form of the signal and the peak of its Fourier transform at $\omega = \omega_0$ should be noted. Adapted from Nielsen and Chuang (2000).

---

[10] More rigorously, ω is an *angular frequency*, measured in rad/s, whereas the *frequency* $\nu = \omega/2\pi$ is measured in Hz. Since we shall use ω almost exclusively, we shall refer to it somewhat casually as the frequency.

where k_B is the Boltzmann constant, $k_B = 1.38 \times 10^{-23}$ J/K. At the ambient temperature for an NMR of 600 MHz the population difference

$$p_0 - p_1 \simeq \frac{\hbar\omega_0}{2k_B T}$$

between the levels $|0\rangle$ and $|1\rangle$ is $\sim 5 \times 10^{-5}$.

The application at time $t = 0$ of a radiofrequency field $\vec{B}_1(t)$ during a time t such that $\omega_1 t = \pi$ with frequency ω_1 lying near the resonance frequency ω_0, that is, a π pulse, makes the spins of the state $|0\rangle$ go to the state $|1\rangle$ and vice versa, resulting in a *population inversion* with respect to the equilibrium populations, so that the sample is out of equilibrium. The return to equilibrium is characterized by a relaxation time [11] T_1, the *longitudinal relaxation time*. In practice, a $\pi/2$ pulse is used: $\omega_1 t = \pi/2$. This corresponds geometrically to rotating the spin by an angle $\pi/2$ about an axis of the xOy plane (Exercise 3.5.1). If the spin is initially parallel to $\vec{B}_0$, it ends up in a plane perpendicular to $\vec{B}_0$, a transverse plane (whereas a π pulse takes the spin to the longitudinal direction $-\vec{B}_0$). The return to equilibrium is then governed by a relaxation time T_2, the *transverse relaxation time*. The time T_1 is of the order of a second and $T_2 \lesssim T_1$; generally, $T_2 \ll T_1$. In any case, the return to equilibrium occurs with the emission of electromagnetic radiation of frequency $\simeq \omega_0$, and Fourier analysis of the signal gives a frequency spectrum which permits the structure of the molecule in question to be reconstructed. This is done on the basis of the following properties.

- The resonance frequency depends on the nuclei through γ.
- For a given nucleus the resonance frequency is slightly modified by the chemical environment of the atom to which the nucleus belongs, and this can be taken into account by defining an effective magnetic field B_0' acting on the nucleus:

$$B_0' = (1 - \sigma)B_0, \qquad \sigma \sim 10^{-6},$$

  where σ is called the *chemical shift*. There are strong correlations between σ and the nature of the chemical grouping to which the nucleus in question belongs.
- The interactions between neighboring nuclear spins provoke a splitting of the resonance frequencies into several subfrequencies which are also characteristic of the chemical groupings.

This is summarized in Fig. 3.6, where a typical NMR spectrum is given. It is important to observe that an NMR measurement has nothing to do with a projective measurement, as defined in Section 2.4. In fact, the NMR signal is a *collective* signal built up by spins located on $\sim 10^{18}$ molecules. When returning

---

[11] When a field $\vec{B}_0$ is applied, thermodynamical equilibrium (3.33) is not established instantaneously, but only after a time $\sim T_1$.

to equilibrium, these spins build up a a macroscopic polarization which precesses about the constant field $\vec{B}_0$. This precession induces an emf in a solenoid (the same solenoid which served to bring the spins to nonequilibrium), and this emf can be measured by standard methods. This gives rise to the free induction signal (FID) schematized in Fig. 3.5. This FID is Fourier analyzed, which allows one to determine the resonance frequencies, as in Fig. 3.6. The reason why the NMR measurement is a purely classical one is that spontaneous emission from a spin in an excited state, which must be described in a quantum framework, is completely negligible, so that the NMR measurement is best described in classical terms.

In the case of magnetic resonance imaging (MRI), it is only the protons contained in water and fats which are of interest. The sample is placed in a nonuniform field $\vec{B}_0$, which makes the resonance frequency depend on the spatial point. Since the signal amplitude is directly proportional to the spin density and therefore to the proton density, by complex computer calculations it is possible to deduce a three-dimensional image of the density of water in biological tissues. At present the spatial resolution is of the order of a millimeter, and an image can be made in 0.1 s. This has allowed the development of functional MRI (fMRI), which can be used, for example, to watch the brain in action by measuring local variations of the blood flow. The longitudinal and transverse relaxation times T_1 and T_2 play a major role in obtaining and interpreting MRI signals.

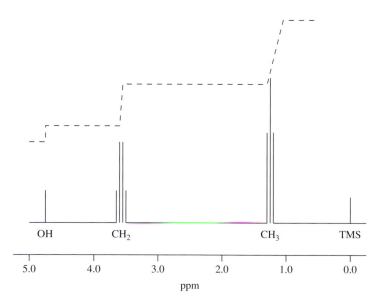

Figure 3.6 NMR spectrum of protons of ethanol CH_3CH_2OH obtained using an NMR of 200 MHz. The three peaks associated with the three groupings OH, CH_3, and CH_2 are clearly seen. The dashed line represents the integrated area of the signals. TMS (tetramethylsilane) is a reference signal.

3.5 Exercises

3.5.1 Rotation operator for spin 1/2

1. Show that the expectation value $\langle \vec{\sigma} \rangle$ of the operator $\vec{\sigma}$ in the state (3.4) is given by $\langle \vec{\sigma} \rangle = \hat{n}$, where $\hat{n}$ is defined in (3.11).

2. Show that

$$\exp\left(-i\frac{\theta}{2}\vec{\sigma}\cdot\hat{p}\right) = I\cos\frac{\theta}{2} - i(\vec{\sigma}\cdot\hat{p})\sin\frac{\theta}{2},$$

where $\hat{p}$ is a unit vector. Hint: calculate $(\vec{\sigma}\cdot\hat{p})^2$. The operator $\exp(-i\theta\vec{\sigma}\cdot\hat{p}/2)$ is the unitary operator $U[\mathcal{R}_{\hat{p}}(\theta)]$ which rotates by an angle θ about the $\hat{p}$ axis. To see this, use the vector $\hat{p} = (-\sin\phi, \cos\phi, 0)$ as the rotation axis and show that a rotation by an angle θ about this axis takes the axis Oz to the vector $\hat{n}$ (3.5). Show that $\exp(-i\theta\vec{\sigma}\cdot\hat{p}/2)|0\rangle$ is just the vector $|\varphi\rangle$ (3.4), the eigenvector of $\vec{\sigma}\cdot\hat{n}$ with eigenvalue $+1$ up to a global phase. What is $\exp(-i\theta\vec{\sigma}\cdot\hat{p}/2)|1\rangle$?

3. When $\phi = -\pi/2$ the rotation is about Ox. Give the explicit matrix form of $U[\mathcal{R}_x(\theta)]$. Comparing with (3.31), show that under the action of $\vec{B}_1(t)$ the state vector rotates by an angle $\theta = -\omega_1 t$ if this field is applied during a time interval $[0, t]$.

3.5.2 Rabi oscillations away from resonance

1. In the nonresonant case, show that starting from (3.30) we obtain a second-order differential equation for $\hat{\lambda}(t)$:

$$\frac{2}{\omega_1}\frac{d^2\hat{\lambda}}{dt^2} - \frac{2i}{\omega_1}\delta\frac{d\hat{\lambda}}{dt} + \frac{1}{2}\omega_1\hat{\lambda} = 0, \qquad \delta = \omega - \omega_0, \tag{3.34}$$

the solutions of which have the form

$$\hat{\lambda}(t) = e^{i\Omega_\pm t}.$$

Show that the values of $\Omega_\pm$ are the roots of a second-order equation and are given as a function of the frequency $\Omega = (\omega_1^2 + \delta^2)^{1/2}$ by

$$\Omega_\pm = \frac{1}{2}\left[\delta \pm \Omega\right].$$

2. The solution of (3.34) for $\hat{\lambda}$ is a linear combination of $\exp(i\Omega_+ t)$ and $\exp(i\Omega_- t)$:

$$\hat{\lambda}(t) = a\exp(i\Omega_+ t) + b\exp(i\Omega_- t).$$

Choose the initial conditions $\hat{\lambda}(0) = 1$, $\hat{\mu}(0) = 0$. Since $\hat{\mu}(0) \propto d\hat{\lambda}(0)/dt$, find a and b as functions of Ω and $\Omega_\pm$.

3. Show that the final result can be written as (see Exercise 6.5.1 for a more elegant proof of this result)

$$\hat{\lambda}(t) = \frac{e^{i\delta t/2}}{\Omega}\left[\Omega\cos\frac{\Omega t}{2} - i\delta\sin\frac{\Omega t}{2}\right],$$

$$\hat{\mu}(t) = \frac{i\omega_1}{\Omega}e^{-i\delta t/2}\sin\frac{\Omega t}{2},$$

which reduces to (3.31) when $\delta = 0$. Starting at $t = 0$ from the state $|0\rangle$, what is the probability of finding the spin in the state $|1\rangle$ at time t? Show that the maximum probability p_-^{max} of making a transition from the state $|0\rangle$ to the state $|1\rangle$ for $\Omega t/2 = \pi/2$ is given by a *resonance curve* of width δ:

$$p_-^{max} = \frac{\omega_1^2}{\omega_1^2 + \delta^2} = \frac{\omega_1^2}{\omega_1^2 + (\omega - \omega_0)^2}.$$

Sketch the curve for p_-^{max} as a function of ω. As shown in Fig. 3.4, the Rabi oscillations are maximal at resonance and decrease rapidly in amplitude with growing δ.

3.6 Further reading

The principles of quantum mechanics are discussed by, for example, Le Bellac (2006), Chapter 4, and Nielsen and Chuang (2000), Chapter 2. The Stern–Gerlach experiment is described in detail by Cohen-Tannoudji *et al.* (1977), Chapter IV, NMR and MRI by Levitt (2001).

4

Quantum correlations

One might expect that going from a single qubit to two qubits would not lead to much of anything new. However, we shall see that the two-qubit structure is extraordinarily rich, because it introduces quantum correlations between the two qubits, correlations which cannot be reproduced using classical probabilistic arguments. Going then from two qubits to n qubits does not lead to anything fundamentally new. As we shall see in Chapter 5, these configurations of quantum systems, called entangled states, are what lead to the special features of quantum computing, due to the exponential growth of the number of states.

4.1 Two-qubit states

The mathematical construction of a two-qubit state rests on the idea of the tensor product, an idea which we shall introduce by means of an elementary example. Let $\mathcal{H}_A$ be a two-dimensional vector space of functions $f_A(x)$ with, for example, the basis vectors $\{\cos x, \sin x\}$:

$$f_A(x) = \lambda_A \cos x + \mu_A \sin x,$$

and let $\mathcal{H}_B$ be another two-dimensional vector space of functions $f_B(y)$ with the basis vectors $\{\cos y, \sin y\}$:

$$f_B(x) = \lambda_B \cos y + \mu_B \sin y.$$

We can construct a function of two variables called the "tensor product of f_A and f_B":

$$f_A(x)f_B(y) = \lambda_A \lambda_B \cos x \cos y + \lambda_A \mu_B \cos x \sin y$$

$$+ \mu_A \lambda_B \sin x \cos y + \mu_A \mu_B \sin x \sin y.$$

A possible basis of the tensor product space is

$$\{\cos x \cos y, \cos x \sin y, \sin x \cos y, \sin x \sin y\}.$$

Any function in this space can be decomposed on this basis:

$$g(x, y) = \alpha \cos x \cos y + \beta \cos x \sin y + \gamma \sin x \cos y + \delta \sin x \sin y,$$

but this function will not in general take the form of the tensor product $f_A(x) f_B(y)$. A necessary (and sufficient) condition for it to take that form is $\alpha\delta = \beta\gamma$.

Let us follow this procedure to construct a two-qubit state mathematically. The first qubit A lives in a Hilbert space $\mathcal{H}_A$ which has orthonormal basis $\{|0_A\rangle, |1_A\rangle\}$, and the second qubit B lives in a Hilbert space $\mathcal{H}_B$ which has orthonormal basis $\{|0_B\rangle, |1_B\rangle\}$. It is natural to represent a physical state in which the first qubit is in the state $|0_A\rangle$ and the second is in the state $|0_B\rangle$ by a vector written as $|X_{00}\rangle = |0_A \otimes 0_B\rangle$. Taking into account all the other possible values of the qubits, we will *a priori* have four possibilities:

$$|X_{00}\rangle = |0_A \otimes 0_B\rangle, \quad |X_{01}\rangle = |0_A \otimes 1_B\rangle, \quad |X_{10}\rangle = |1_A \otimes 0_B\rangle, \quad |X_{11}\rangle = |1_A \otimes 1_B\rangle. \tag{4.1}$$

The notation $\otimes$ stands for the tensor product. It is not difficult to construct a state in which the qubit A is in the normalized state

$$|\varphi_A\rangle = \lambda_A|0_A\rangle + \mu_A|1_A\rangle, \qquad |\lambda_A|^2 + |\mu_A|^2 = 1,$$

and the qubit B is in the normalized state

$$|\varphi_B\rangle = \lambda_B|0_B\rangle + \mu_B|1_B\rangle, \qquad |\lambda_B|^2 + |\mu_B|^2 = 1.$$

We shall denote this state as $|\varphi_A \otimes \varphi_B\rangle$:

$$\begin{aligned}|\varphi_A \otimes \varphi_B\rangle &= \lambda_A\lambda_B|0_A \otimes 0_B\rangle + \lambda_A\mu_B|0_A \otimes 1_B\rangle \\ &\quad + \mu_A\lambda_B|1_A \otimes 0_B\rangle + \mu_A\mu_B|1_A \otimes 1_B\rangle \\ &= \lambda_A\lambda_B|X_{00}\rangle + \lambda_A\mu_B|X_{01}\rangle + \mu_A\lambda_B|X_{10}\rangle + \mu_A\mu_B|X_{11}\rangle.\end{aligned} \tag{4.2}$$

The correspondence with the preceding functional space is obvious. We have constructed the space $\mathcal{H}_A \otimes \mathcal{H}_B$ as the *tensor product* of the spaces $\mathcal{H}_A$ and $\mathcal{H}_B$. We note that the vector $|\varphi_A \otimes \varphi_B\rangle$ is also normalized.[1] Physicists are rather lax in their notation, and following in this tradition the reader will sometimes here find $|\varphi_A \otimes \varphi_B\rangle$, or $|\varphi_A\rangle \otimes |\varphi_B\rangle$, or even $|\varphi_A\varphi_B\rangle$, with the symbol for the tensor product omitted.

The crucial point is that the most general state of $\mathcal{H}_A \otimes \mathcal{H}_B$ is not of the form of a tensor product $|\varphi_A \otimes \varphi_B\rangle$; states of the form $|\varphi_A \otimes \varphi_B\rangle$ make up only a small subset (not even a subspace!) of the vectors of $\mathcal{H}_A \otimes \mathcal{H}_B$. The most general state has the form

$$\begin{aligned}|\Psi\rangle &= \alpha_{00}|0_A \otimes 0_B\rangle + \alpha_{01}|0_A \otimes 1_B\rangle + \alpha_{10}|1_A \otimes 0_B\rangle + \alpha_{11}|1_A \otimes 1_B\rangle \\ &= \alpha_{00}|X_{00}\rangle + \alpha_{01}|X_{01}\rangle + \alpha_{10}|X_{10}\rangle + \alpha_{11}|X_{11}\rangle,\end{aligned} \tag{4.3}$$

---

[1] More rigorously, it should be checked that the product $|\varphi_A \otimes \varphi_B\rangle$ is independent of the choice of bases in $\mathcal{H}_A$ and $\mathcal{H}_B$. This can be proved immediately; see Exercise 4.6.1.

and for $|\Psi\rangle$ to be of the form $|\varphi_A \otimes \varphi_B\rangle$ a necessary (and sufficient) condition is that

$$\alpha_{00}\alpha_{11} = \alpha_{01}\alpha_{10},$$

which *a priori* has no reason to be valid. Let us give a very simple example of a state $|\Phi\rangle$ which is *not* of the form $|\varphi_A \otimes \varphi_B\rangle$:

$$|\Phi\rangle = \frac{1}{\sqrt{2}}\left(|0_A \otimes 1_B\rangle + |1_A \otimes 0_B\rangle\right). \qquad (4.4)$$

Here

$$\alpha_{00} = \alpha_{11} = 0, \qquad \alpha_{01} = \alpha_{10} = \frac{1}{\sqrt{2}},$$

and $\alpha_{00}\alpha_{11} \neq \alpha_{01}\alpha_{10}$. We also define the tensor product $M_A \otimes M_B$ of two operators M_A and M_B as

$$[M_A \otimes M_B]_{i_A p_B; j_A q_B} = [M_A]_{i_A j_A}[M_B]_{p_B q_B}.$$

As an example, let us give the tensor product of two 2×2 matrices:

$$M_A = \begin{pmatrix} a & b \\ c & d \end{pmatrix}, \qquad M_B = \begin{pmatrix} \alpha & \beta \\ \gamma & \delta \end{pmatrix}.$$

The matrix $M_A \otimes M_B$ is a 4×4 matrix, with the lines and columns ordered as 00, 01, 10, 11:

$$M_A \otimes M_B = \begin{pmatrix} aM_B & bM_B \\ cM_B & dM_B \end{pmatrix} = \begin{pmatrix} a\alpha & a\beta & b\alpha & b\beta \\ a\gamma & a\delta & b\gamma & b\delta \\ c\alpha & c\beta & d\alpha & d\beta \\ c\gamma & c\delta & d\gamma & d\delta \end{pmatrix}.$$

A two-qubit state which does not have the form $|\varphi_A \otimes \varphi_B\rangle$ is called an *entangled state*. The *fundamental* property of such a state is the following: if $|\Psi\rangle$ is an entangled state, then the qubit A cannot be in a definite quantum state $|\varphi_A\rangle$. Let us first show this for a special case, that of the state $|\Phi\rangle$ (4.4). Let M be a physical property of the qubit A. In the space $\mathcal{H}_A \otimes \mathcal{H}_B$ this physical property is represented by $M \otimes I_B$. We calculate its expectation value $\langle\Phi|M\Phi\rangle$ as

$$\langle M\rangle_\Phi = \langle\Phi|M\Phi\rangle = \frac{1}{2}\left[\langle 0_A \otimes 1_B| + \langle 1_A \otimes 0_B|\right]\left[|(M0_A) \otimes 1_B\rangle + |(M1_A) \otimes 0_B\rangle\right]$$

$$= \frac{1}{2}\left(\langle 0_A|M0_A\rangle + \langle 1_A|M1_A\rangle\right), \qquad (4.5)$$

where we have used

$$\langle 0_B|0_B\rangle = \langle 1_B|1_B\rangle = 1, \qquad \langle 0_B|1_B\rangle = \langle 1_B|0_B\rangle = 0.$$

Let us prove that there is no state

$$|\varphi_A\rangle = \lambda|0_A\rangle + \mu|1_A\rangle$$

such that

$$\langle\Phi|M\Phi\rangle = \langle\varphi_A|M\varphi_A\rangle.$$

Computing the expectation value of M, we obtain

$$\langle\varphi_A|M\varphi_A\rangle = |\lambda|^2\langle0_A|M0_A\rangle + (\lambda^*\mu\langle0_A|M1_A\rangle + \lambda\mu^*\langle1_A|M0_A\rangle) + |\mu|^2\langle1_A|M1_A\rangle.$$

A necessary condition for reproducing (4.5) would be $|\lambda| = |\mu| = 1/\sqrt{2}$, but then the terms involving $\lambda^*\mu$ would not vanish (unless $\langle0_A|M1_A\rangle$ vanishes accidentally), in contradiction with (4.5). The result (4.5) has a simple physical interpretation: the state of the qubit A is an *incoherent* mixture of 50% of the state $|0_A\rangle$ and 50% of the state $|1_A\rangle$, and not a linear superposition. In summary, it is not possible in general to describe *part* of a quantum system by a state vector.

An example of an incoherent superposition is natural or unpolarized light. It is an incoherent mixture of 50% light polarized along Ox and 50% light polarized along Oy, whereas light polarized at 45° is a *coherent* superposition of 50% light polarized along Ox and 50% light polarized along Oy:

$$|\theta = \pi/4\rangle = \frac{1}{\sqrt{2}}(|x\rangle + |y\rangle).$$

Right-handed circularly polarized light is also a coherent superposition:

$$|R\rangle = \frac{1}{\sqrt{2}}(|x\rangle + \mathrm{i}|y\rangle).$$

We see the importance of phases: for example, the states $|\theta = \pi/4\rangle$ and $|R\rangle$ both correspond to 50% probability of observing a photon polarized along Ox or along Oy, but these two states are completely different: one is a linear polarization and the other is a circular polarization.

Box 4.1: Example of a physical realization of an entangled state

Obtaining an entangled state starting from a tensor product is not completely straightforward. It is necessary to introduce an interaction between the two qubits. Let us take as an example two spins 1/2. A possible interaction [2] between these two spins is

$$\hat{H} = \frac{\hbar\omega}{2}\vec{\sigma}_A \cdot \vec{\sigma}_B.$$

---

[2] Such an interaction might originate in the interaction between the two magnetic moments associated with the spins, but in general it will more likely be associated with an exchange interaction originating in the Pauli exclusion principle.

We use the result of Exercise 4.6.4

$$\frac{1}{2}(I + \vec{\sigma}_A \cdot \vec{\sigma}_B)|ij\rangle = |ji\rangle$$

to show that

$$(\vec{\sigma}_A \cdot \vec{\sigma}_B)\frac{1}{\sqrt{2}}(|10\rangle + |01\rangle) = (\vec{\sigma}_A \cdot \vec{\sigma}_B)|\Phi_+\rangle = |\Phi_+\rangle,$$

$$(\vec{\sigma}_A \cdot \vec{\sigma}_B)\frac{1}{\sqrt{2}}(|10\rangle - |01\rangle) = (\vec{\sigma}_A \cdot \vec{\sigma}_B)|\Phi_-\rangle = -3|\Phi_-\rangle.$$

The vectors $|\Phi_+\rangle$ and $|\Phi_-\rangle$ are eigenvectors of $\vec{\sigma}_A \cdot \vec{\sigma}_B$ with the eigenvalues [3] $+1$ and -3, respectively. Let us start at time $t = 0$ from a nonentangled state, for example, $|\Phi(t = 0)\rangle = |10\rangle$. To obtain its time evolution it is sufficient to decompose this state on $|\Phi_+\rangle$ and $|\Phi_-\rangle$:

$$|\Phi(t = 0)\rangle = \frac{1}{\sqrt{2}}(|\Phi_+\rangle + |\Phi_-\rangle).$$

We can immediately write down the time evolution:

$$e^{-i\hat{H}t/\hbar}|\Phi(0)\rangle = \frac{1}{\sqrt{2}}\left(e^{-i\omega t/2}|\Phi_+\rangle + e^{3i\omega t/2}|\Phi_+\rangle\right)$$

$$= \frac{1}{\sqrt{2}}e^{i\omega t/2}\left(e^{-i\omega t}|\Phi_+\rangle + e^{i\omega t}|\Phi_+\rangle\right)$$

$$= e^{i\omega t/2}\left(\cos\omega t|10\rangle - i\sin\omega t|01\rangle\right).$$

One can now choose $\omega t = \pi/4$ to obtain the entangled state $|\Psi\rangle$:

$$|\Psi\rangle = \frac{1}{\sqrt{2}}(|10\rangle - i|01\rangle).$$

The difficulty is that $\hat{H}$ is in general an interaction *internal* to the system, which, in contrast to the external interactions used to manipulate the individual qubits, cannot easily be turned on and off in order to adjust t. If the interaction is a short-range one, it is possible to move the two qubits closer together and farther apart in order to control the time over which they interact. The construction of entangled states in the case of NMR using series of radiofrequency pulses will be discussed in Section 6.1. In the case of trapped ions, a two-ion state is entangled by allowing it to pass through the intermediary of an ion vibrational mode (Section 6.2). It is also possible to obtain an entangled state of two objects by using a third auxiliary object, for example, two atoms can be entangled by making them interact with a photon of a resonant cavity.

---

[3] Physicists will recognize these as corresponding to the triplet ($|\Phi_+\rangle$) and singlet ($|\Phi_-\rangle$) states.

4.2 The state operator (or density operator)

Now let us generalize these results to a quantum system formed of any two subsystems, where we use $|i_A\rangle(|i_B\rangle)$ to refer to an orthonormal basis of the subsystem $A(B)$. To simplify the notation, it will be convenient to make the substitutions $i_A \to i$ and $i_B \to \mu$. The most general state then is

$$|\Phi\rangle = \sum_{i,\mu} \alpha_{i\mu} |i \otimes \mu\rangle. \tag{4.6}$$

Let M be a physical property of the subsystem A:

$$|M\Phi\rangle = \sum_{i,\mu} \alpha_{i\mu} |Mi \otimes \mu\rangle.$$

We calculate the expectation value of M as

$$
\begin{aligned}
\langle \Phi|M\Phi \rangle &= \sum_{j,\nu} \sum_{i,\mu} \alpha^*_{j\nu} \alpha_{i\mu} \langle j \otimes \nu | Mi \otimes \mu \rangle \\
&= \sum_{i,j} \sum_{\mu} \alpha^*_{j\mu} \alpha_{i\mu} \langle j|Mi\rangle = \sum_{i,j} \rho_{ij} \langle j|Mi\rangle = \sum_{i,j} \rho_{ij} M_{ji} = \mathrm{Tr}(\rho M),
\end{aligned}
\tag{4.7}
$$

where $\mathrm{Tr}A$ stands for the *trace* $\sum_i A_{ii}$ of an operator A, that is, the sum of its diagonal elements. It is straightforward to prove that $\mathrm{Tr}AB = \mathrm{Tr}BA$, from which one deduces that the trace is basis independent. In obtaining (4.7) we have used

$$\langle j \otimes \nu | Mi \otimes \mu \rangle = \delta_{\nu\mu} \langle j|Mi\rangle,$$

because in $\mathcal{H}_A \otimes \mathcal{H}_B$, M is actually $M \otimes I_B$. The equation (4.7) defines an object which will play a crucial role, the *state operator* (or density operator)[4] ρ of the subsystem A:

$$\boxed{\rho_{ij} = \sum_\mu \alpha_{i\mu} \alpha^*_{j\mu}} \tag{4.8}$$

The state operator of the subsystem A is called the *reduced state operator* and is denoted ρ_A. The subsystem A is not in general described by a state vector, but by a state operator which allows us to compute the expectation values of physical properties. This state operator is Hermitian ($\rho = \rho^\dagger$), positive[5] ($\rho \geq 0$ as is easily proved from (4.8)), and it has unit trace $\mathrm{Tr}\rho = 1$:

$$\mathrm{Tr}\rho = \sum_i \rho_{ii} = \sum_i \sum_\mu |\alpha_{i\mu}|^2 = ||\Phi||^2 = 1.$$

---

[4] The standard terminology is "density operator." However, this historical term is completely unjustified: to what density does it refer? We prefer the term "state operator," which is the generalization to mixtures of the term "state vector" for pure states.

[5] A positive (or nonnegative) operator A is one for which $\langle \varphi|A\varphi\rangle$ is real and $\langle \varphi|A\varphi\rangle \geq 0 \ \forall|\varphi\rangle$ (it is strictly positive if $\langle \varphi|A\varphi\rangle > 0$). It is necessarily Hermitian in a complex space. A necessary and sufficient condition for a Hermitian operator to be positive is that its eigenvalues be nonnegative.

Physical states such as those studied in Chapter 2 are called *pure states*: they are described by a state vector. It is easy to check that the state operator of a pure state obeys $\rho^2 = \rho$ and vice versa: any state operator satisfying $\rho^2 = \rho$ describes a pure state (Exercise 4.6.2). However, the most general description of a quantum system must be given in terms of a state operator.

Since ρ is Hermitian, it can be diagonalized and written in an orthonormal basis $|i\rangle$ as

$$\rho = \sum_i \mathsf{p}_i |i\rangle\langle i|. \tag{4.9}$$

Since ρ is positive $\mathsf{p}_i \geq 0$, and the condition $\mathrm{Tr}\rho = 1$ gives $\sum_i \mathsf{p}_i = 1$, so that the p_i can be interpreted as probabilities. It can be said that ρ represents a *statistical mixture* (or simply a *mixture*) of states $|i\rangle$, each state $|i\rangle$ having a probability p_i; in the preparation stage, each state $|i\rangle$ is prepared with a probability p_i, without any phase coherence between the different states $|i\rangle$.

It is easy to generalize (4.8) when a quantum system (AB) is described by a state operator ρ_{AB} with matrix elements[6] $\rho^{AB}_{i\mu;j\nu}$, and not by a state vector. Let M be a physical property of the system A, which is therefore represented in the space $\mathcal{H}_A \otimes \mathcal{H}_B$ by the Hermitian operator $M \otimes I_B$. We wish to find an operator ρ_A such that the expectation value of M is given by

$$\langle M \rangle = \mathrm{Tr}(\rho_A M). \tag{4.10}$$

Using the same argument as above, we calculate the expectation value of $M \otimes I_B$:

$$\langle M \otimes I_B \rangle = \mathrm{Tr}_{AB}\left(\rho_{AB}[M \otimes I_B]\right)$$
$$= \sum_{ij\mu\nu} \rho^{AB}_{i\mu;j\nu} M_{ji}\delta_{\nu\mu} = \sum_{i,j} M_{ji} \sum_{\mu} \rho^{AB}_{i\mu;j\mu}. \tag{4.11}$$

The expression generalizing (4.8) then shows that ρ_A has the form

$$\boxed{\rho^A_{ij} = \sum_\mu \rho^{AB}_{i\mu;j\mu}, \qquad \rho_A = \mathrm{Tr}_B \rho_{AB}} \tag{4.12}$$

because the expectation value of M is given by (4.10) with the choice (4.12) for ρ_A. It can be shown that (4.12) is the unique solution giving the correct expectation value of M. The operation which takes us from ρ_{AB} to ρ_A is called the *partial trace* of ρ_{AB} with respect to B.

The importance of the concept of state operator is confirmed by the *Gleason theorem*, which we shall state without proof and which basically says that the most general description of a quantum system is given by a state operator.

---

[6] To make the notation more readable, AB is written as a superscript to make room for the subscripts labeling the matrix elements.

The Gleason theorem Let a set of projectors $\mathcal{P}_i$ act on a Hilbert space of states $\mathcal{H}$ and let there be a test associated with each $\mathcal{P}_i$ where the probability $\mathsf{p}(\mathcal{P}_i)$ of passing the test satisfies

$$0 \leq \mathsf{p}(\mathcal{P}_i) \leq 1, \qquad \mathsf{p}(I) = 1,$$

as well as

$$\mathsf{p}(\mathcal{P}_i \cup \mathcal{P}_j) = \mathsf{p}(\mathcal{P}_i) + \mathsf{p}(\mathcal{P}_j) \text{ if } \mathcal{P}_i \cap \mathcal{P}_j = \emptyset \, (\text{or } \mathcal{P}_i \mathcal{P}_j = \delta_{ij} \mathcal{P}_i).$$

This property should hold for any set of mutually orthogonal $\mathcal{P}_i$ such that $\sum_i \mathcal{P}_i = I$. Then if the dimension of $\mathcal{H} \geq 3$, there exists a positive Hermitian operator ρ of unit trace such that

$$\mathsf{p}(\mathcal{P}_i) = \mathrm{Tr}(\rho \mathcal{P}_i).$$

In other words, if we wish to associate a probability $\mathsf{p}(\mathcal{P}_i)$ with an ensemble of tests $\mathcal{P}_i$ which has "reasonable" properties, then this probability will be given by a trace involving a state operator.

If $|\Phi\rangle$ is a tensor product of the form $|\varphi_A \otimes \varphi_B\rangle$ and if to $|\Phi\rangle$ we apply a unitary transformation which is a tensor product of transformations acting on A and B, $U_A \otimes U_B$, this corresponds simply to a change of orthonormal basis in the spaces $\mathcal{H}_A$ and $\mathcal{H}_B$ and an entangled state cannot be made. To make an entangled state, *it is necessary to make the two qubits interact*. In contrast to the superposition of two states which is a basis dependent concept, entanglement is a basis independent concept. The Schmidt purification theorem, whose proof is left to Exercise 4.6.5, allows us to give more precise statements.

The Schmidt purification theorem Any state $|\Phi\rangle$ of $\mathcal{H}_A \otimes \mathcal{H}_B$ can be written as

$$|\Phi\rangle = \sum_i \sqrt{\mathsf{p}_i} \, |i_A \otimes i_B\rangle \qquad (4.13)$$

with

$$\langle i_A | j_A \rangle = \langle i_B | j_B \rangle = \delta_{ij}.$$

The states $|i_A\rangle$ and $|i_B\rangle$ clearly depend on $|\Phi\rangle$. This expression immediately gives the reduced state operators ρ_A and ρ_B. To show this, let us begin with the full state operator ρ_{AB}:

$$\rho_{AB} = |\Phi\rangle\langle\Phi| = \sum_{i,j} |i_A \otimes i_B\rangle\langle j_A \otimes j_B|.$$

Let $|i\rangle$ be an orthonormal basis of $\mathcal{H}$. It is easy to calculate the traces using the following result:

$$\mathrm{Tr}|\varphi\rangle\langle\psi| = \sum_i \langle i|\varphi\rangle\langle\psi|i\rangle = \sum_i \langle\psi|i\rangle\langle i|\varphi\rangle = \langle\psi|\varphi\rangle, \qquad (4.14)$$

because $\sum_i |i\rangle\langle i| = I$ and so the state operators ρ_A and ρ_B are given by

$$\rho_A = \sum_i \mathsf{p}_i |i_A\rangle\langle i_A|, \qquad \rho_B = \sum_i \mathsf{p}_i |i_B\rangle\langle i_B| \qquad (4.15)$$

with the *same* p_i. The number of nonzero p_i is the *Schmidt number*. If we apply to the state $|\Phi\rangle$ a unitary transformation which is the tensor product of transformations acting on A and B, $U_A \otimes U_B$, we cannot change the Schmidt number by manipulating the qubits A and B *separately*. We recover the above result for a tensor product by noting that the Schmidt number of a tensor product is 1. If $\mathcal{H}_A$ and $\mathcal{H}_B$ have dimension N, a state of the form

$$|\Phi\rangle = \frac{1}{\sqrt{N}} \sum_{i=1}^{N} e^{i\alpha(i)} |i_A \otimes i_B\rangle, \qquad (4.16)$$

where $\exp[i\alpha(i)]$ is a phase factor, is termed a *maximally entangled state*, or a *Bell state*. For example, $|\Phi\rangle$ in (4.4) is a Bell state. The reduced state matrices corresponding to (4.16) are multiples of the identity: $\rho_A = \rho_B = I/N$. This is a characteristic property of maximally entangled states.

4.3 The quantum no-cloning theorem

The indispensable condition for the quantum cryptography method of Section 2.5 to be perfectly secure is that the spy Eve not be able to reproduce (clone) the state of the particle sent by Alice to Bob while leaving Bob's measurement result unchanged, so that interception of the message is undetectable. The impossibility of Eve doing this is guaranteed by the quantum no-cloning theorem. To demonstrate this theorem, we suppose that we wish to duplicate an *unknown* quantum state $|\chi_1\rangle$. Of course, if $|\chi_1\rangle$ were known, there would be no problem because the preparation procedure would be known. The system on which we wish to print the copy is denoted $|\varphi\rangle$ and is the equivalent of a blank page. For example, if we wish to clone a spin $1/2$ state $|\chi_1\rangle$, then $|\varphi\rangle$ will also be a state of spin $1/2$. The evolution of the state vector in the cloning process must be of the form

$$|\chi_1 \otimes \varphi\rangle \to |\chi_1 \otimes \chi_1\rangle. \qquad (4.17)$$

This evolution is governed by a unitary operator U whose exact form is unimportant:

$$|U(\chi_1 \otimes \varphi)\rangle = |\chi_1 \otimes \chi_1\rangle. \qquad (4.18)$$

This operator U must be universal (because the photocopying operation cannot depend on the state to be copied) and therefore independent of $|\chi_1\rangle$, which is unknown by hypothesis. If we wish to clone a second original $|\chi_2\rangle$ we must have

$$|U(\chi_2 \otimes \varphi)\rangle = |\chi_2 \otimes \chi_2\rangle.$$

Let us now evaluate the scalar product

$$X = \langle \chi_1 \otimes \varphi | U^\dagger U (\chi_2 \otimes \varphi) \rangle$$

in two different ways:

(1) $X = \langle \chi_1 \otimes \varphi | \chi_2 \otimes \varphi \rangle = \langle \chi_1 | \chi_2 \rangle,$

(2) $X = \langle \chi_1 \otimes \chi_1 | \chi_2 \otimes \chi_2 \rangle = (\langle \chi_1 | \chi_2 \rangle)^2.$ (4.19)

The result is that either $|\chi_1\rangle \equiv |\chi_2\rangle$ or $\langle \chi_1 | \chi_2 \rangle = 0$. It is possible to clone a state $|\chi_1\rangle$ or an orthogonal state, but not a linear superposition of the two. This proof of the no-cloning theorem explains why it is not possible in quantum cryptography to restrict oneself to a basis of orthogonal polarization states $\{|x\rangle, |y\rangle\}$ for photons. It is the use of linear superpositions of the polarization states $|x\rangle$ and $|y\rangle$ which allows the presence of a spy to be detected. The no-cloning theorem makes it impossible for Eve to clone the photon sent by Alice to Bob when its polarization is unknown to Bob; if Eve could perform this cloning, she would then be able to produce a large number of such photons and measure the polarization without problem, see Exercise 2.6.1.

4.4 Decoherence

Let us consider two qubits A and B in the entangled state

$$|\Psi\rangle = \lambda |0_A 0_B\rangle + \mu |1_A 1_B\rangle$$

and compute the state matrix of qubit A using (4.14):

$$\rho_A = \text{Tr}_B |\Psi\rangle\langle\Psi| = |\lambda|^2 |0_A\rangle\langle 0_A| + |\mu|^2 |1_A\rangle\langle 1_A| = \begin{pmatrix} |\lambda|^2 & 0 \\ 0 & |\mu|^2 \end{pmatrix}.$$ (4.20)

All the information on the phases of the complex numbers λ and μ seems to have disappeared, and we are left in the $\{|0_A\rangle, |1_A\rangle\}$ basis with a diagonal state matrix. It is easy to generalize the preceding argument and derive the following theorem: *if a pair of states of the system of interest becomes correlated with mutually orthogonal states of another system, then all the phase coherence between the orthogonal states of the first system is lost.* This loss of phase coherence is called *decoherence.* Since the information on the phases is contained in the nondiagonal matrix elements of ρ, these matrix elements are often called *coherences.* However, this theorem should be interpreted with care. First of all, although ρ_A is diagonal in the $\{|0_A\rangle, |1_A\rangle\}$ basis, this will not be the case in general in another basis except if $\rho_A = I/2$. For example, in the $\{|+_A\rangle, |-_A\rangle\}$ basis

$$|\pm_A\rangle = \frac{1}{\sqrt{2}} (|0_A\rangle \pm |1_A\rangle)$$

ρ_A (4.20) takes the form

$$\rho_A = \frac{1}{2} \begin{pmatrix} 1 & |\lambda|^2 - |\mu|^2 \\ |\lambda|^2 - |\mu|^2 & 1 \end{pmatrix}. \tag{4.21}$$

The second remark is that the information on the phases is not really lost: it is only *locally* lost, that is, it is lost if we restrict ourselves to measurements of physical properties of qubit A. Joint physical properties of qubits A and B will depend in general on the phases of λ and μ. In the case of a quantum computer, a large number of qubits become entangled and the state matrix of any individual qubit is almost diagonal. However, if the qubits are perfectly isolated, the global state vector retains the phase information, and indeed this must be so if we want the quantum computation to be meaningful. The third remark is that coherence may be recovered dynamically, even if it seems to have been temporarily lost. To explain this point, let us assume that the states $|0_B\rangle$ and $|1_B\rangle$ are not exactly orthogonal, but that $\langle 0_B | 1_B \rangle = \varepsilon$, $|\varepsilon| \ll 1$, and that the two-qubit Hamiltonian has the form

$$\hat{H} = \frac{\hbar}{2} K \begin{pmatrix} 0 & 0 & 0 & 1 \\ 0 & 0 & 0 & 0 \\ 0 & 0 & 0 & 0 \\ 1 & 0 & 0 & 0 \end{pmatrix} = \frac{\hbar}{2} KX. \tag{4.22}$$

It is easily checked (for example, by relabeling the rows and columns of the matrix X in the order 00, 11, 01, 11) that the evolution operator is

$$\exp(-i\hat{H}t/\hbar) = I_A \otimes I_B \cos\frac{Kt}{2} - iX \sin\frac{Kt}{2}. \tag{4.23}$$

If we start at time $t = 0$ from the state $|\Psi_0\rangle = |0_A 0_B\rangle$, the state vector at time $t = \pi/2K$ will be

$$\Psi\left(t = \frac{\pi}{2K}\right) = \frac{1}{\sqrt{2}}\left(|0_A 0_B\rangle - i|1_A 1_B\rangle\right),$$

and the corresponding state matrix of qubit A becomes

$$\rho_A = \frac{1}{2} \begin{pmatrix} 1 & i\varepsilon \\ -i\varepsilon & 1 \end{pmatrix}. \tag{4.24}$$

It appears that the state matrix exhibits almost perfect decoherence. However, if we wait until time $t = \pi/K$ the state vector becomes $-i|1_A 1_B\rangle$. It is possible to work out a more precise model where the states $|0_B\rangle$ and $|1_B\rangle$ represent almost nonoverlapping, and consequently almost orthogonal, wave packets. It can then be shown that $K \propto \varepsilon$, so that the oscillations predicted by (4.23) have a very long period, and the qubit A may appear to have lost coherence for a long time.

As long as we have complete control over all the quantum variables, we have only "adiabatic" or "false" decoherence: at some instant of time, the measurement of local physical properties may not depend on the phases, but the phases are still there and they reappear in the measurement of more complex physical properties and/or from dynamical evolution. We shall face "true" decoherence if we lose control over some of the quantum variables. For example, suppose that in a Young's slit experiment performed with complex molecules, such as fullerenes in a thermally excited state, a photon is emitted when the molecule passes through one of the slits and escapes to infinity. If the wavelength of this photon is shorter than the distance between the two slits,[7] the state of the photon will be almost orthogonal to that of a photon emitted by the molecule going through the other slit. The path of the molecule will become correlated with orthogonal degrees of freedom of the environment, so that we get information on which path is taken and the interference is destroyed. In this case it is clear that we have lost control over the photon degrees of freedom, and information has leaked into the environment in an uncontrollable fashion. This is an example of true decoherence: the system becomes entangled with orthogonal states of the environment, but we do not have access to these states. It can also be said that "the environment measures the system," since the emitted photon measures the path of the molecule. In general, the environment is a very complicated quantum system, and quantum coherences are distributed over such a large number of degrees of freedom that they become unobservable.

It follows from the preceding discussion that if we want to retain control over the operation of a quantum computer, it is essential that the computer be immune to decoherence: the qubits must not be coupled to the *quantum* degrees of freedom of their environment. In other words, an ideal quantum computer must be completely isolated. In many models the characteristic time which controls the decay of coherence, called the *decoherence time*, is inversely proportional to some positive power of the size of the system, often the square of this size. Thus, we expect decoherence to be more and more important as the number of qubits increases. The following example will illustrate this property. Suppose that the interaction of a qubit with the environment during a time interval Δt has the following effect on a single qubit:

$$|0\rangle \to |0\rangle, \qquad |1\rangle \to -|1\rangle$$

with a probability $p = \Gamma \Delta t \ll 1$. The decoherence time is $\tau_D = 1/\Gamma$. We have not explicitly written out the states of the environment, the only important point being

---

[7] This will happen if the temperature of the molecule is sufficiently high, so that it can be found in a highly excited state.

that the states $|0\rangle$ and $|1\rangle$ do not become entangled with states of the environment. If the qubit is in the state

$$|\psi\rangle = \frac{1}{\sqrt{2}}(|0\rangle + |1\rangle),$$

then the interaction will transform it into

$$|\psi'\rangle = \frac{1}{\sqrt{2}}(|0\rangle - |1\rangle) = \sigma_z|4\rangle \tag{4.25}$$

with probability p, and the initial phase relation between the two components of $|\psi\rangle$ will be lost, introducing errors in the computation. The process (4.25) is conventionally called *phase flip*. Now, consider the n-qubit state

$$|\Psi_n\rangle = \frac{1}{\sqrt{2}}(|00\cdots0\rangle + |11\cdots1\rangle). \tag{4.26}$$

The phase relation between the two components of $|\Psi\rangle$ will be lost as soon as *one* of the qubits flips sign. It is reasonable to assume that each of the qubits interacts with the environment independently of the others. Then in the time interval Δt the state $|\Psi\rangle$ will be transformed into

$$|\Psi\rangle = \frac{1}{\sqrt{2}}(|00\cdots0\rangle - |11\cdots1\rangle)$$

with probability $\mathsf{p}_n = n\Gamma\Delta t$. In other words, the decoherence time for the system of n qubits will be shorter by a factor of n compared to the decoherence time for a single qubit: $\tau_D(n) = \tau_D/n$.

   To conclude this section, let us describe a simple model for the coupling of a qubit to its environment which leads to phase decoherence. This model is conventionally called the *phase damping channel*. In this model the state of the qubit does not change, but the environment, which is initially in the state $|0_E\rangle$, is sent with probability p into the state $|1_E\rangle(|2_E\rangle)$ if the qubit is in the state $|0_A\rangle(|1_A\rangle)$:

$$|0_A0_E\rangle \to \sqrt{1-\mathsf{p}}|0_A0_E\rangle + \sqrt{\mathsf{p}}|0_A1_E\rangle = |0_A\rangle \otimes \left(\sqrt{1-\mathsf{p}}\,|0_E\rangle + \sqrt{\mathsf{p}}\,|1_E\rangle\right),$$

$$|1_A0_E\rangle \to \sqrt{1-\mathsf{p}}|1_A0_E\rangle + \sqrt{\mathsf{p}}|1_A2_E\rangle = |1_A\rangle \otimes \left(\sqrt{1-\mathsf{p}}\,|0_E\rangle + \sqrt{\mathsf{p}}\,|2_E\rangle\right). \tag{4.27}$$

   One can imagine, for example, that the qubit elastically scatters a photon of the cosmic microwave background, and that the final state of the photon depends on the state of the qubit. We note that the states $|0_A\rangle$ and $|1_A\rangle$ do not become entangled with the environment, while any linear combination of these two states would become entangled. States which do not become entangled with their environment are termed *pointer states*. The most general initial state is

$$|\Phi\rangle = (\lambda|0_A\rangle + \mu|1_A\rangle) \otimes |0_E\rangle , \qquad (4.28)$$

so that the initial state matrix of the qubit is

$$\rho_A = \begin{pmatrix} |\lambda|^2 & \lambda\mu^* \\ \lambda^*\mu & |\mu|^2 \end{pmatrix} = \begin{pmatrix} \rho_{00} & \rho_{01} \\ \rho_{10} & \rho_{11} \end{pmatrix}. \qquad (4.29)$$

The process (4.27) can be represented in the Hilbert space $\mathcal{H}_A \otimes \mathcal{H}_E$ by a unitary operator U, which is only partially known, [8] and

$$U|\Phi\rangle = \lambda\sqrt{1-\mathsf{p}}\,|0_A 0_E\rangle + \lambda\sqrt{\mathsf{p}}\,|0_A 1_E\rangle + \mu\sqrt{1-\mathsf{p}}\,|1_A 0_E\rangle + \mu\sqrt{\mathsf{p}}\,|1_A 2_E\rangle. \quad (4.30)$$

Using (4.14), it is now straightforward to find the transformed state matrix ρ'_A of the qubit: [9]

$$\rho'_A = \mathrm{Tr}_E\left[U|\Phi\rangle\langle\Phi|U^\dagger\right] = |\lambda|^2|0_A\rangle\langle 0_A| $$
$$+ |\mu|^2|1_A\rangle\langle 1_A| + (\lambda\mu^*(1-\mathsf{p})|0_A\rangle\langle 1_A| + \mathrm{H.c.}) \qquad (4.31)$$

or

$$\rho'_A = \begin{pmatrix} \rho_{00} & (1-\mathsf{p})\rho_{01} \\ (1-\mathsf{p})\rho_{10} & \rho_{11} \end{pmatrix}. \qquad (4.32)$$

After n iterations of (4.30) we find

$$\rho'^{(n)}_A = \begin{pmatrix} \rho_{00} & (1-\mathsf{p})^n\rho_{01} \\ (1-\mathsf{p})^n\rho_{10} & \rho_{11} \end{pmatrix} \xrightarrow[n\to\infty]{} \begin{pmatrix} \rho_{00} & \rho_{01}e^{-\Gamma t} \\ \rho_{10}e^{-\Gamma t} & \rho_{11} \end{pmatrix}. \qquad (4.33)$$

Indeed, if we assume that p is proportional to Δt, $\mathsf{p} = \Gamma\Delta t$, and we observe the qubit during a time interval t, $n = t/\Delta t$, then

$$\rho_{01}(t) = \rho_{01}(1-\Gamma\Delta t)^{t/\Delta t} \xrightarrow[\Delta t\to 0]{} \rho_{01}e^{-\Gamma t}. \qquad (4.34)$$

The initial state decays into an incoherent mixture of the states $|0_A\rangle$ and $|1_A\rangle$ with a decoherence time of $\tau_D = 1/\Gamma$. For $t \to \infty$, the state matrix becomes diagonal

$$\rho(t) \xrightarrow[t\to\infty]{} \begin{pmatrix} \rho_{00} & 0 \\ 0 & \rho_{11} \end{pmatrix}.$$

---

[8] U is a 6×6 matrix and only four of its matrix elements are given in (4.27), but the missing entries can be filled in while preserving the unitarity.
[9] For example, $\mathrm{Tr}_E\left(|0_A 1_E\rangle\langle 1_A 1_E|\right) = |0_A\rangle\langle 1_A|$.

It is essential to observe that no unitary evolution in $\mathcal{H}_A$ can lead from the initial state matrix ρ_A (4.29) to the final diagonal from. Indeed, a unitary transformation transforms a pure state into a pure state, and there is no unitary operator such that

$$U\rho_A U^\dagger = \begin{pmatrix} |\lambda|^2 & 0 \\ 0 & |\mu|^2 \end{pmatrix}.$$

The unitary evolution takes place in the $\mathcal{H}_A \otimes \mathcal{H}_B$ Hilbert space.

4.5 The Bell inequalities

One proof [10] of the nonclassical nature of the correlations of an entangled state is given by the Bell inequalities, which we shall explain using an example. Let us suppose that we have constructed pairs of photons A and B traveling in opposite directions whose linear polarizations along Ox or Oy are entangled (Fig. 4.1):

$$|\Phi\rangle = \frac{1}{\sqrt{2}} \left(|x_A x_B\rangle + |y_A y_B\rangle \right). \tag{4.35}$$

Alice and Bob are able to measure the polarizations of the photons issued from a single pair, because the photon pairs are separated by a time interval sufficient for them not to overlap. Alice measures the polarization of photon A and Bob the polarization of photon B, then they check to see whether the polarizations are correlated: if Alice and Bob orient *both* of their analyzers either along the axis Ox or along Oy, they can check that the two photons either pass through both

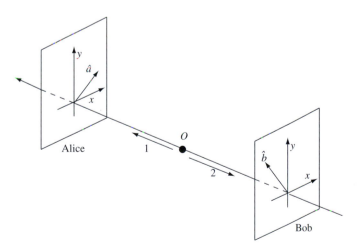

Figure 4.1 Configuration of an EPR type of experiment.

---

[10] This section is a digression from the main topic and may be omitted from a first reading.

their analyzers or are stopped by both of them. Mathematically, this results from the probability amplitudes

$$\langle x_A x_B | \Phi \rangle = \frac{1}{\sqrt{2}}, \qquad \langle x_A y_B | \Phi \rangle = 0, \qquad \langle x_A y_B | \Phi \rangle = 0, \qquad \langle y_A y_B | \Phi \rangle = \frac{1}{\sqrt{2}}.$$

To write this result in a convenient form, it is useful to describe the correlation of the polarizations as follows (A_x and B_x are just the operator $M = \mathcal{P}_x - \mathcal{P}_y$ introduced in Section 2.4):

$$A_x = +1 \text{ if polarization } A \parallel Ox, \qquad B_x = +1 \text{ if polarization } B \parallel Ox,$$

$$A_x = -1 \text{ if polarization } A \parallel Oy, \qquad B_x = -1 \text{ if polarization } B \parallel Oy.$$

Under these conditions, Alice and Bob observe, for example, the following series of results:

$$\text{Alice}: \quad A_x = +--+-+++--,$$

$$\text{Bob}: \quad B_x = +--+-+++--,$$

which gives the expectation value of the product $A_x B_x$:

$$\langle A_x B_x \rangle = 1. \tag{4.36}$$

Upon reflection, this result is not very surprising. It is a variation of the game of the two customs inspectors. [11] Two travelers A and B, each carrying a suitcase, depart in opposite directions from the origin and eventually are checked by two customs inspectors Alice and Bob. One of the suitcases contains a red ball and the other a green ball, but the travelers have picked up their closed suitcases at random and do not know what color the ball inside is. If Alice checks the suitcase of traveler A, she has a 50% chance of finding a green ball. But if in fact she finds a green ball, clearly Bob will find a red ball with 100% probability! Correlations between the two suitcases were introduced at the time of departure, and these correlations reappear as a correlation between the results of Alice and Bob.

However, as first noted by Einstein, Podolsky, and Rosen (EPR) in a celebrated paper [12] (which used a different example, ours being due to Bohm), the situation becomes much less commonplace if Alice and Bob decide to perform another series of measurements using the orientations $\hat{\theta}$ and $\hat{\theta}_\perp$ instead of Ox and Oy. In fact, $|\Phi\rangle$ is invariant under rotation about Oz, because (2.19) can be used to show immediately (Exercise 4.6.8) that $|\Phi\rangle$ can also be written as

$$|\Phi\rangle = \frac{1}{\sqrt{2}} \left(|\theta_A \theta_B\rangle + |\theta_{\perp A} \theta_{\perp B}\rangle \right). \tag{4.37}$$

---

[11] Invented just for this occasion!

[12] Einstein *et al.* (1935). The term "EPR paradox" is sometimes used, but in fact there is nothing paradoxical in the EPR analysis.

If A_x is replaced by A_θ, then

$$A_\theta = +1 \text{ if polarization } A \parallel \hat{\theta}, \qquad B_\theta = +1 \text{ if polarization } B \parallel \hat{\theta},$$

$$A_\theta = -1 \text{ if polarization } A \parallel \hat{\theta}_\perp, \qquad B_\theta = -1 \text{ if polarization } B \parallel \hat{\theta}_\perp.$$

Then as in (4.16) we will have

$$\langle A_\theta B_\theta \rangle = 1. \tag{4.38}$$

Knowing the polarization of photon A along $\hat{\theta}$, we can predict with certainty the polarization of photon B along $\hat{\theta}$ (or $\hat{\theta}_\perp$) for any choice of θ. One gets the impression that Alice and Bob can communicate instantaneously, even if they are separated by several light years, and thus that relativity is violated. Of course, this is only an illusion, because in order to be able to compare their results and check (4.38), Alice and Bob must be able to exchange messages via a classical path and therefore at a speed less than that of light. Moreover, it is straightforward to reproduce these correlations using a classical model (Fig. 4.2), in which the correlations are fixed in advance.

However, this will no longer be possible if Alice and Bob decide to use different axes $\hat{a}$ and $\hat{b}$. We use the following generalization of the case of parallel axes (see the example in Fig. 4.2): polarization $\parallel \hat{a} : A(\hat{a}) = +1, \ldots,$ polarization

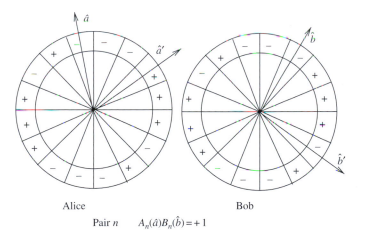

Alice   Bob

Pair n $A_n(\hat{a})B_n(\hat{b}) = +1$

Figure 4.2 A classical model for EPR correlations. The suitcases of travelers A and B are now circles divided into small angular sectors defining the orientations $\hat{a}, \ldots, \hat{b}'$ in the plane xOy, which are labeled $+$ for polarization in the given direction or $-$ for polarization in the orthogonal direction. The two circles are identical and two diametrically opposite points are identified and both labeled $+$ or $-$. The figure corresponds to $A_{\hat{a}} = -1$, $A_{\hat{a}'} = +1$, $B_{\hat{b}} = -1$, and $B_{\hat{b}'} = -1$.

$\perp \hat{b}: B(\hat{b}) = -1$. Then we construct the expectation value $E(\hat{a}, \hat{b})$ measured in N experiments with $N \to \infty$:

$$E(\hat{a}, \hat{b}) = \lim_{N \to \infty} \frac{1}{N} \sum_{n=1}^{N} A_n(\hat{a}) B_n(\hat{b}). \tag{4.39}$$

Let us now construct the combination X_n with the orientations ($\hat{a}$ or $\hat{a}'$) for a and ($\hat{b}$ or $\hat{b}'$) for b, $A_n = A_n(\hat{a})$, $B_n' = B_n(\hat{b}')$, ..., where n numbers the pairs, and Alice and Bob are able to identify unambiguously the photons belonging to the same pair:

$$X_n = A_n B_n + A_n B_n' + A_n' B_n' - A_n' B_n = A_n(B_n + B_n') + A_n'(B_n' - B_n). \tag{4.40}$$

Here $X_n = \pm 2$, which leads to the following *Bell inequality*:

$$|\langle X \rangle| = \left| \lim_{N \to \infty} \frac{1}{N} \sum_{n=1}^{N} X_n \right| \leq 2. \tag{4.41}$$

The quantity X_n is "counterfactual" because it cannot be measured for a single pair: there are four possible choices for the orientation of the measurement axes, but only one choice for a particular pair. The EPR point of view is that *each photon carries all the information on its intrinsic polarization* and that the four combinations $A_n B_n \cdots A_n' B_n'$ exist for any pair n, even if only one can be measured in a given experiment. However, this does not necessarily mean that the EPR viewpoint is incorrect, because, as Feynman has stated, "It is not true that we can pursue science completely by using only those concepts which are directly subject to experiment." Proof that the EPR viewpoint is incorrect will come from experiment.

What does quantum physics actually say? It is easy to calculate $E(\hat{a}, \hat{b})$. Owing to the rotational invariance, it is always possible to choose $\hat{a}$ parallel to Ox. We write $|\Phi\rangle$ as

$$|\Phi\rangle = \frac{1}{\sqrt{2}} \left[|x_A\rangle \left(\cos\theta |\theta_B\rangle - \sin\theta |\theta_{\perp B}\rangle \right) + |y_A\rangle \left(\sin\theta |\theta_B\rangle + \cos\theta |\theta_{\perp B}\rangle \right) \right],$$

writing out $|x_B\rangle$ and $|y_B\rangle$ as functions of $|\theta_B\rangle$ and $|\theta_{\perp B}\rangle$ (see (2.19)). We can then immediately calculate the scalar products:

$$\langle x_A \theta_B | \Phi \rangle = \frac{1}{\sqrt{2}} \cos\theta, \qquad \langle x_A \theta_{\perp B} | \Phi \rangle = -\frac{1}{\sqrt{2}} \sin\theta,$$

$$\langle y_A \theta_B | \Phi \rangle = \frac{1}{\sqrt{2}} \sin\theta, \qquad \langle y_A \theta_{\perp B} | \Phi \rangle = \frac{1}{\sqrt{2}} \cos\theta,$$

and so

$$E(\hat{x}, \hat{\theta}) = \frac{1}{2} \left[2\cos^2\theta - 2\sin^2\theta \right] = \cos(2\theta), \tag{4.42}$$

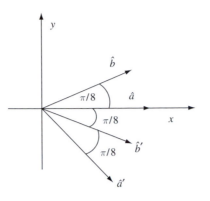

Figure 4.3 An optimal angular configuration.

or, in a form which is manifestly rotationally invariant,

$$E(\hat{a}, \hat{b}) = \cos(2\hat{a} \cdot \hat{b}).$$

With the angles chosen as in Fig. 4.3 we find

$$|\langle X \rangle| = 2\sqrt{2} \simeq 2.82. \tag{4.43}$$

There is no classical correlation which can reproduce the quantum correlations: *the quantum correlations are too strong to be explained classically.* Even if the qubits A and B are several light years apart, they cannot be considered as separate entities and there is no local probabilistic classical algorithm which is capable of reproducing their correlations. The qubits A and B form a unique entity; they are nonseparable, or entangled.

Let us also note that the no-cloning theorem forbids propagation of information at superluminal velocities. Alice can choose to use either the basis $\{|x\rangle, |y\rangle\}$ or the basis $\{|+\pi/4\rangle, |-\pi/4\rangle\}$ to measure the polarization of her photon. If Bob could clone his own photon, he would be able to measure its polarization and instantly deduce the basis chosen by Alice for her corresponding photon, even if she were located several light years away from him.

4.6 Exercises

4.6.1 Basis independence of the tensor product

Let us suppose that we have constructed the tensor product of two spaces $\mathcal{H}_A$ and $\mathcal{H}_B$ starting from the bases $\{|m_A\rangle\}$ and $\{|n_B\rangle\}$:

$$|\varphi_A \otimes \chi_B\rangle = \sum_{m,n} c_m d_n |m_A \otimes n_B\rangle.$$

Let $|i_A\rangle$ and $|j_B\rangle$ be two other orthonormal bases of $\mathcal{H}_A$ and $\mathcal{H}_B$ deduced from the bases $|m_A\rangle$ and $|n_B\rangle$ by the unitary transformations R $(R^{-1} = R^\dagger)$ and S $(S^{-1} = S^\dagger)$, respectively:

$$|i_A\rangle = \sum_m R_{im} |m_A\rangle, \qquad |j_B\rangle = \sum_n S_{jn} |n_B\rangle.$$

Calculate the tensor product $|i \otimes j\rangle$. To construct the tensor product, we now decompose $|\varphi\rangle$ and $|\chi\rangle$ in the respective bases $|i\rangle$ and $|j\rangle$:

$$|\varphi\rangle = \sum_{i=1}^{N} \hat{c}_i |i_A\rangle, \qquad |\chi\rangle = \sum_{j=1}^{M} \hat{d}_j |j_B\rangle.$$

Show that

$$\sum_{i,j} \hat{c}_i \hat{d}_j |i_A \otimes j_B\rangle = |\varphi \otimes \chi\rangle.$$

4.6.2 Properties of the state operator

1. Starting from (4.9),

$$\rho = \sum_i \mathsf{p}_i |i\rangle\langle i|, \qquad \sum_i \mathsf{p}_i = 1,$$

show that the most general state operator ρ must possess the following properties.

1. It must be Hermitian: $\rho = \rho^\dagger$.
2. It must have unit trace: $\mathrm{Tr}\rho = 1$.
3. It must be positive: $\langle \varphi|\rho|\varphi\rangle \geq 0 \ \forall|\varphi\rangle$.

Show that the expectation value of a physical property M is

$$\langle M\rangle = \mathrm{Tr}(\rho M).$$

2. Show also that if $\rho^2 = \rho$, then all the p_i are zero except one, which is equal to unity, and prove that the condition $\rho^2 = \rho$ is the necessary and sufficient condition for a state to be pure. Also show that $\mathrm{Tr}\rho^2 = 1$ is a necessary and sufficient condition for the state operator to describe a pure state.

4.6.3 The state operator for a qubit and the Bloch vector

1. We wish to find the most general form of ρ for a qubit; ρ will be represented by a 2×2 state matrix. Show that the most general Hermitian matrix of unit trace in $\mathcal{H}$ has the form

$$\rho = \begin{pmatrix} a & c \\ c^* & 1-a \end{pmatrix},$$

where a is a real number and c is a complex number. Show that the positivity of the eigenvalues of ρ introduces a supplementary constraint on the matrix elements:

$$0 \le a(1-a) - |c|^2 \le \frac{1}{4}.$$

Show that the necessary and sufficient condition for the quantum state described by ρ to be represented by a vector of $\mathcal{H}$ is $a(1-a) = |c|^2$. Calculate a and c for the matrix ρ describing the normalized state vector $|\psi\rangle = \lambda|0\rangle + \mu|1\rangle$ with $|\lambda|^2 + |\mu|^2 = 1$, and show that in this case $a(1-a) = |c|^2$.

2. Show that ρ can be written as a function of a vector $\vec{b}$ called the *Bloch vector*:

$$\rho = \frac{1}{2}\begin{pmatrix} 1+b_z & b_x - ib_y \\ b_x + ib_y & 1 - b_z \end{pmatrix} = \frac{1}{2}\left(I + \vec{b}\cdot\vec{\sigma}\right),$$

provided that $|\vec{b}|^2 \le 1$. Show that a quantum state represented by a vector of $\mathcal{H}$ corresponds to the case $|\vec{b}|^2 = 1$. To interpret the vector $\vec{b}$ physically, we calculate the expectation value of $\vec{\sigma}$:

$$\langle\sigma_i\rangle = \text{Tr}\,(\rho\,\sigma_i).$$

Show that $\vec{b}$ is the expectation value of $\vec{\sigma}$.

3. When the spin is placed in a constant magnetic field $\vec{B}$, the Hamiltonian is given by

$$H = -\frac{1}{2}\,\gamma\,\vec{\sigma}\cdot\vec{B},$$

where γ is a constant. Assuming that $\vec{B}$ is parallel to the axis Oz, $\vec{B} = (0, 0, B)$, write down the evolution equation for ρ and show that the vector $\vec{b}$ rotates (precesses) about $\vec{B}$ with an angular frequency to be determined.

4.6.4 The SWAP operator

1. Show that the operator

$$\frac{1}{2}\left(I + \vec{\sigma}_A\cdot\vec{\sigma}_B\right)$$

permutes the values of the two bits A and B:

$$\frac{1}{2}\left(I + \vec{\sigma}_A\cdot\vec{\sigma}_B\right)|i_A j_B\rangle = |j_A i_B\rangle.$$

The notation $\vec{\sigma}_A\cdot\vec{\sigma}_B$ stands for both the scalar product and the tensor product.

2. The operator $\frac{1}{2}\left(I + \vec{\sigma}_A \cdot \vec{\sigma}_B\right)$ is called the *SWAP* operator. Its matrix representation in the basis $\{|00\rangle, |01\rangle, |10\rangle, |11\rangle\}$ is

$$U_{\text{SWAP}} = \begin{pmatrix} 1 & 0 & 0 & 0 \\ 0 & 0 & 1 & 0 \\ 0 & 1 & 0 & 0 \\ 0 & 0 & 0 & 1 \end{pmatrix}.$$

Check that its square root $U_{\text{SWAP}}^{1/2}$ is given by

$$U_{\text{SWAP}}^{1/2} = \frac{1}{1+i} \begin{pmatrix} 1+i & 0 & 0 & 0 \\ 0 & 1 & i & 0 \\ 0 & i & 1 & 0 \\ 0 & 0 & 0 & 1+i \end{pmatrix},$$

and that the so-called cZ gate can be constructed from

$$\text{cZ} = e^{i\pi\sigma_z^A/4} \, e^{-i\pi\sigma_z^B/4} \, U_{\text{SWAP}}^{1/2} \, e^{i\pi\sigma_z^A/2} \, U_{\text{SWAP}}^{1/2} = \begin{pmatrix} I & 0 \\ 0 & \sigma_z \end{pmatrix}.$$

4.6.5 The Schmidt purification theorem

Let $|\varphi_{AB}\rangle \in \mathcal{H}_A \otimes \mathcal{H}_B$ be a pure state of the system AB, and let $\{|m_A\rangle\}$ and $\{|\mu_B\rangle\}$ be two orthonormal bases of $\mathcal{H}_A$ and $\mathcal{H}_B$. The most general decomposition of $|\varphi_{AB}\rangle$ on the basis $\{|m_A \otimes \mu_B\rangle\}$ of $\mathcal{H}_A \otimes \mathcal{H}_B$ is written as

$$|\varphi_{AB}\rangle = \sum_{m,\mu} c_{m\mu} |m_A \otimes \mu_B\rangle.$$

We define the vectors $|\tilde{m}_B\rangle \in \mathcal{H}_B$ as

$$|\tilde{m}_B\rangle = \sum_{\mu} c_{m\mu} |\mu_B\rangle$$

and rewrite the above decomposition as

$$|\varphi_{AB}\rangle = \sum_{m} |m_A \otimes \tilde{m}_B\rangle.$$

The vectors $\{|\tilde{m}_B\rangle\}$ do not *a priori* form an orthonormal basis of $\mathcal{H}_B$. We choose as the basis of $\mathcal{H}_A$ a set of vectors $\{|m_A\rangle\}$ which diagonalizes ρ_A:

$$\rho_A = \text{Tr}_B |\varphi_{AB}\rangle\langle\varphi_{AB}| = \sum_{m} p_m |m_A\rangle\langle m_A|.$$

Comparing this expression for ρ_A with

$$\rho_A = \sum_{m,n} \langle \tilde{n}_B | \tilde{m}_B \rangle | m_A \rangle \langle n_A |,$$

prove that

$$\langle \tilde{n}_B | \tilde{m}_B \rangle = \mathsf{p}_m \delta_{mn}$$

and the vectors $|\tilde{n}_B\rangle$ are orthogonal after all. How can an orthonormal basis $|n_B\rangle$ be constructed? How should the terms such that $\mathsf{p}_n = 0$ be treated? Show that in this basis

$$|\varphi_{AB}\rangle = \sum_n \mathsf{p}_n^{1/2} |n_A \otimes n_B\rangle.$$

4.6.6 A model for phase damping

Let us consider the NMR case where a spin 1/2 is submitted to a fluctuating magnetic field $\vec{B}_0(t)$. The state $|1\rangle$ can assumed to be stable (spontaneous emission is negligible), but the resonance frequency $\omega_0 = \gamma B_0/\hbar$ is time dependent. The state vector of the spin system at time t is

$$|\Psi(t)\rangle = \lambda(t)|0\rangle + \mu(t)|1\rangle,$$

with $\lambda(t)$ and $\mu(t)$ given by

$$\mathrm{i}\dot{\lambda}(t) = -\frac{1}{2}\omega_0(t)\lambda(t), \qquad \mathrm{i}\dot{\mu}(t) = \frac{1}{2}\omega_0(t)\mu(t), \qquad \lambda(0) = \lambda_0, \mu(0) = \mu_0.$$

The solution is

$$\lambda(t) = \lambda_0 \exp\left(\frac{\mathrm{i}}{2}\int_0^t \omega_0(t')\,\mathrm{d}t'\right), \qquad \mu(t) = \mu_0 \exp\left(-\frac{\mathrm{i}}{2}\int_0^t \omega_0(t')\,\mathrm{d}t'\right).$$

Assume that $\omega_0(t)$ is a Gaussian stationary random function with connected autocorrelation function

$$C(t') = \langle \omega_0(t+t')\omega_0(t)\rangle - \langle \omega_0\rangle^2,$$

where $\langle \bullet \rangle$ is an ensemble average over all realizations of the random function. Also assume that

$$C(t') \simeq C \exp\left(-\frac{|t'|}{\tau}\right).$$

Show that the populations ρ_{00} and ρ_{11} are time independent, but that the time evolution of the coherences is given by

$$\rho_{01}(t) = \rho_{01}(t=0)\mathrm{e}^{\mathrm{i}\langle \omega_0\rangle t}\,\mathrm{e}^{-C\tau t}, \qquad t \gg \tau.$$

4.6.7 Amplitude damping channel

In the so-called *amplitude damping channel*, we have instead of (4.27) the following evolution

$$U|0_A \otimes 0_E\rangle = |0_A \otimes 0_E\rangle,$$

$$U|1_A \otimes 0_E\rangle = \sqrt{1-p}\,|1_A \otimes 0_E\rangle + \sqrt{p}\,|0_A \otimes 1_E\rangle.$$

This is a model for describing the spontaneous decay of an atom in an excited state $|1_A\rangle$ into its ground state $|0_A\rangle$, while $|0_E\rangle$ is a state with zero photons and $|1_E\rangle$ a state with one photon. The probability of decay during a time interval Δt is p.

1. Starting from the state

$$|\Phi\rangle = (\lambda|0_A\rangle + \mu|1_A\rangle) \otimes |0_E\rangle,$$

compute the final state matrix ρ'_A. Show that the time evolution may be written in the form

$$\rho_A(t=0) \rightarrow \rho(t) = \begin{pmatrix} 1 - e^{-\Gamma t}\rho_{11} & e^{-\Gamma t/2}\rho_{01} \\ e^{-\Gamma t/2}\rho_{10} & e^{-\Gamma t}\rho_{11} \end{pmatrix}.$$

Deduce from this that, in this model, the transverse relaxation time T_2 is twice the longitudinal relaxation time T_1, $T_2 = 2T_1$ (see Section 3.4).

2. Suppose that at time Δt one observes the environment in the zero photon state $|0_E\rangle$. What is then the state of the atom? Show that the failure to detect a photon has changed the state of the atom.

4.6.8 Invariance of the state (4.35) under rotation

Using

$$|\theta\rangle = \cos\theta|x\rangle + \sin\theta|y\rangle,$$

$$|\theta_\perp\rangle = -\sin\theta|x\rangle + \cos\theta|y\rangle,$$

show that

$$|\Phi\rangle = \frac{1}{\sqrt{2}} (|x_A x_B\rangle + |y_A y_B\rangle) = \frac{1}{\sqrt{2}} (|\theta_A \theta_B\rangle + |\theta_{A\perp} \theta_{B\perp}\rangle).$$

4.7 Further reading

A popularized approach to entangled states can be found in Hey and Walters (2003), Chapter 8. The state operator is studied in Nielsen and Chuang (2000), Chapter 2, Preskill (1999), Chapter 3, and Le Bellac (2006), Chapter 6. Very clear accounts of decoherence are found in Leggett (2002), Zurek (1991) and in Paz and Zurek (2002). Aspect (1999) reviews the experimental tests of Bell inequalities. Advanced theoretical discussions are found in Mermin (1993) and in Peres (1993), Chapters 6 and 7. A proof of the Gleason theorem and a demonstration of Schmidt decomposition are given by Peres (1993), Chapters 5 and 7. Interference using complex molecules is discussed by Arndt *et al.* (2005).

5

Introduction to quantum computing

5.1 General remarks

It is easy to represent integers in terms of qubits in the same manner as for ordinary bits. Let us suppose that we wish to write an integer between 0 and 7 in a register of qubits. If this were a classical register, we would need 3 bits. In a system of base 2, a number between 0 and 7 can be represented in binary notation as a sequence of three digits 0 or 1. A classical register will store *one* of the 8 following configurations:

$$0 = \{000\}, \qquad 1 = \{001\}, \qquad 2 = \{010\}, \qquad 3 = \{011\},$$

$$4 = \{100\}, \qquad 5 = \{101\}, \qquad 6 = \{110\}, \qquad 7 = \{111\}.$$

A system of three qubits will also allow a number from 0 to 7 to be stored, for example, by making these numbers correspond to the following 8 states of three qubits:

$$0 : |000\rangle, \qquad 1 : |001\rangle, \qquad 2 : |010\rangle, \qquad 3 : |011\rangle,$$

$$4 : |100\rangle, \qquad 5 : |101\rangle, \qquad 6 : |110\rangle, \qquad 7 : |111\rangle. \tag{5.1}$$

Here we have omitted the tensor product notation; for example, $|101\rangle$ is abbreviated notation for $|1_A \otimes 0_B \otimes 1_C\rangle$, where the qubits A, B, and C have their state vector in $\mathcal{H}_A$, $\mathcal{H}_B$, and $\mathcal{H}_C$, respectively. We use $|x\rangle$, $x = 0, \ldots, 7$, to denote one of the eight states of (5.1), for example, $|5\rangle = |101\rangle$. It is not difficult to generalize to the case of n qubits; representing a number less than $N = 2^n$ requires n qubits, and $|x\rangle$ denotes the state vector with

$$0 \leq x \leq 2^n - 1.$$

The basis of the Hilbert space $\mathcal{H}^{\otimes n}$ formed using orthonormal vectors $|x\rangle$ is called the *computational basis*. Since we can construct a linear superposition of the eight states (5.1), it can be concluded that the state vector of a system of three spins allows us to store $2^3 = 8$ numbers at the same time, while if n spins are used we can store 2^n numbers! However, if, for example, spins 1/2 are used for the physical support of the qubits, a measurement of the three spins along the axis Oz will necessarily give one of the eight states (5.1). We have at our disposal important virtual information, but when we try to materialize it in a measurement we can do no better than for a classical system: the measurement gives one of eight numbers, and not all eight at the same time! It is therefore necessary to go further in order truly to exploit the possibilities of a quantum computer and find algorithms which are specific to it. This will be explained later on in this chapter, and for now we shall give only a schematic description of the principle by which such a quantum computer functions.

A calculation performed on a quantum computer is shown schematically in Fig. 5.1, where n qubits are all prepared in the state $|0\rangle$ at time $t = t_0$. This is the preparation stage of the quantum system, and the initial state vector belongs to a Hilbert space of 2^n dimensions, $\mathcal{H}^{\otimes n}$. This initialization stage is not a unitary operation, but a projective measurement, and it is a dissipative process. The qubits then undergo a unitary quantum evolution described by a unitary operator $U(t, t_0)$ acting in $\mathcal{H}^{\otimes n}$ which performs the desired operations, for example, the calculation of a function. The experimental difficulty is to avoid any interaction with the environment, because then the phenomenon of decoherence would make the evolution nonunitary. As we have seen in Section 4.4, if there is an interaction with the environment, the unitary evolution occurs in a Hilbert space which is larger than $\mathcal{H}^{\otimes n}$, because it includes the degrees of freedom of the environment along with those of the qubits. Interactions with external classical fields are compatible

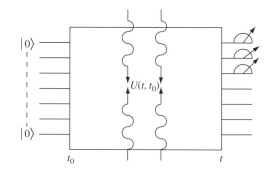

Figure 5.1 Schematic depiction of the basic principle of a quantum calculation. n qubits are prepared in the state $|0\rangle$. They undergo a unitary and deterministic evolution in the space $\mathcal{H}^{\otimes n}$ from time $t = t_0$ to time t described by a unitary operator $U(t, t_0)$ acting in $\mathcal{H}^{\otimes n}$. The wiggly arrows represent interactions with external classical fields. A measurement of the qubits (or a subset of the qubits, the first three in this figure) is made at time t. The diagrams are read from left to right, in the direction opposite to that of the operator products.

with unitary evolution and they are needed to manipulate qubits by Rabi oscillations. [1] Once the quantum evolution has been completed, a measurement is made on the qubits (or on a subset of qubits) at time t in order to obtain the result of the calculation. An important point is that *the state of the calculation cannot be observed* between t_0 and t, because any measurement would modify the unitary evolution: the box $U(t, t_0)$ of Fig. 5.1 is a black box which must not be tampered with. The qubits are measured at the entrance and at the exit of the box, but not inside it. Another essential point is that the unitary evolution is *reversible*: [2] if we know the state vector at time t, we can recover the state vector at time t_0 using $U^{-1}(t, t_0) = U(t_0, t)$.

5.2 Reversible calculation

The passage of the initial qubit state at $t = t_0$ to the final qubit state at t occurs via a reversible operation, and the algorithms of a quantum computer must necessarily be reversible. This is not the case with the algorithms used on classical computers, which are irreversible, and so the latter cannot be transposed directly to quantum computers. Most of the usual logic gates are irreversible, because they correspond to a transformation (2 bits $\rightarrow$ 1 bit), and the final state of a single bit does not allow the reconstruction of the initial two-bit state. For example, the NAND gate

$$x \uparrow y = 1 \oplus xy,$$

where $\oplus$ is mod 2 addition, gives the correspondence

$$(00) \rightarrow 1, \qquad (01) \rightarrow 1, \qquad (10) \rightarrow 1, \qquad (11) \rightarrow 0,$$

and knowledge of the final state does not permit reconstruction of the initial state. It is known that the NAND gate and the COPY operation are sufficient for constructing any logic circuit. An interesting question is whether or not all the usual logic operations can be performed reversibly on a classical computer.

---

[1] A note for physicists: the reason why the action of a classical field is compatible with unitary evolution is subtle, see Leggett (2002). Let us consider the states $|0\rangle$ and $|1\rangle$ of a spin 1/2 in a uniform magnetic field parallel to Oz. If the spin is initially in the excited state $|1\rangle$, then application of a π-pulse will send it in the ground state $|0\rangle$, and the external magnetic field will gain one photon. So, it appears that the transition has left a mark in the environment, a feature which should lead to decoherence. However, the (quantum) state of the field is a coherent state containing a very large number of photons and this state changes in a negligible manner when one adds one photon: the spin does not become entangled with the (quantum) state of the electromagnetic field. Of course, the energy difference between the two levels must be small enough, so that spontaneous emission is negligible. Otherwise the spin would also interact with the vacuum fluctuations of the *quantized* electromagnetic field, and its evolution would be no longer be unitary, see Exercise 4.6.7. All this has been confirmed in a beautiful experiment using neutron interferometry by Badurek *et al.* (1985); this experiment is thoroughly analyzed in Omnès (1994), Chapter 11.

[2] A second note for physicists: reversible evolution and invariance under time reversal should not be confused, as time reversal is represented in $\mathcal{H}$ by an anti-unitary operator, whereas $U^{-1}(t, t_0) = U(t_0, t)$ is unitary.

This question was initially only of theoretical interest, and was first raised by Landauer and Bennett, who wondered if it were possible to perform a calculation without energy dissipation. In fact, in spite of its abstract nature, information is necessarily carried by some physical support. [3] As a bonus, in following this line of inquiry Bennett was finally (after more than a century!) able to obtain a satisfactory solution of the paradox of the Maxwell demon (see Box 5.1). According to Landauer, a calculation involving irreversible operations like the loss of a bit of information in a NAND operation costs a thermodynamical entropy of at least $k_B \ln 2$ per bit, where k_B is the Boltzmann constant ($k_B = 1.38 \times 10^{-23}$ J/K), and therefore leads to the dissipation of an energy $\Delta E = k_B T \ln 2$ into the environment, where T is the absolute temperature of the computer. At present the problem is academic, because for an actual PC the energy dissipated per erased bit is already $\Delta E \sim 500 k_B T$ simply owing to electricity consumption, and so we are nowhere near $k_B T$. However, it is possible that this question will become of practical import some time in the future. The energy dissipated per logical operation has decreased by ten orders of magnitude in the past 50 years, so that it might be that the $k_B T$ limit becomes relevant in a little more than 10 or 15 years.

The real reason for the interest in reversible calculation is the possibility of transposing classical algorithms to quantum computing. As we have already mentioned, direct transposition is impossible, because quantum computing is reversible, and so the NAND operation must be replaced by an equivalent reversible operation. It is also necessary to find the equivalent of the COPY operation without coming into conflict with the no-cloning theorem when transposition is made to the quantum version of the gates. This can be done using two logic gates, the cNOT gate and the Toffoli gate (Fig. 5.2). If the bits entering the *control-NOT (cNOT)*

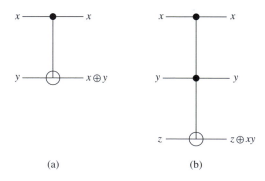

(a)   (b)

Figure 5.2 The cNOT gate (a) and the Toffoli gate (b). The black points represent the control bits and the circles represent the target bits.

---

[3] "Information is physical," according to Landauer, who went as far as deducing (debatably in the author's opinion) from this that mathematics and information science are branches of physics!

gate are (x, y), where x is the *control bit* and y is the *target bit*, the action of the cNOT gate on the target bit depends on the state of the control bit according to the scheme

$$\text{cNOT}: \quad (x, y) \rightarrow (x, x \oplus y). \tag{5.2}$$

Box 5.1: Maxwell's demon and the physical nature of information

In this box we show that it is impossible to ignore the fact that information must be carried by a physical support; otherwise, we come into conflict with the second law of thermodynamics. In 1871 Maxwell imagined the following device. A container filled with a gas at absolute temperature T is divided into two compartments of identical volume, separated by a wall in which a small hole is pierced (Fig. 5.3). A demon can observe the velocity of the molecules and open and close this hole by a door without expending energy. The molecules in the container have an average velocity of several hundred meters per second at ambient temperature ($T \simeq 300\,\text{K}$),

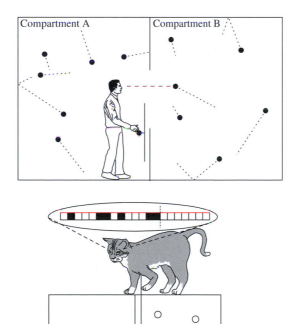

Figure 5.3 Maxwell's demon. The demon stores the position of the molecules in his memory.

but some are faster and others are slower. The demon opens the door when he sees a fast molecule traveling from the left-hand to the right-hand compartment, and also when he sees a slow molecule traveling from the right-hand to the left-hand compartment. Therefore, the average speed of the molecules in the right-hand compartment will increase, while that of the molecules in the left-hand compartment will decrease, with the total energy of the gas remaining constant. Since the average velocity is related to T and to the mass m of a molecule as

$$v \simeq \sqrt{\frac{k_B T}{m}},$$

the right-hand compartment will become warmer than the left-hand one. These two compartments can then be used as two heat sources at different temperatures to run a heat engine, thus making it possible to obtain work from only a single heat source, contradicting the second law of thermodynamics (equivalently, we could make a refrigerator without a motor, which is also forbidden by the second law).

In 1929, this problem was reduced to its bare essentials by Szilard, who considered a gas limited to a single molecule. This molecule can be localized in one or the other compartment without expending energy, and it does work by pushing a piston until it occupies the entire container, while taking energy from the outside in the form of heat. The expansion is done at constant temperature, and the work done is given by

$$W_0 = k_B T \int_{V/2}^{V} \frac{\mathrm{d}V'}{V'} = k_B T \ln 2,$$

where V is the volume of the container. The operation can be performed n times so as to obtain an arbitrarily large amount of work $W = nW_0 = nk_B T \ln 2$, all using a single heat source.

The paradox was elucidated by Bennett in 1982. He noted that this device *does not function on a cycle*, which is the condition for the second law of thermodynamics to be valid, because the localization of the molecule in one or the other compartment during the n operations involves the assumption that this information will be stored in a memory of n bits. If we wish to erase the contents of this memory in order to restart from zero and perform a complete cycle, this would release into the environment an entropy of at least $nk_B \ln 2$, and therefore dissipate into the environment an energy of at least $nk_B T \ln 2$, which would convert all the work performed into heat.

Looking at this in more detail, we see that when the compartment in which the molecule is located has been determined, the information entropy of the system is one bit (corresponding to a thermodynamic entropy $k_B \ln 2$), because the position of the molecule and the contents of the memory are correlated. Once the expansion has occurred, the value of the information entropy is two bits, because the information about the compartment is lost. The information entropy of the environment must therefore decrease by one bit, that is, an energy $k_B T \ln 2$ equal to the work W_0 is supplied by the environment. When the memory is erased, we return to a one-bit entropy for the system, which means that the environment will receive at least one bit, because the entropy of the ensemble (system + environment) can only increase. If the operations are performed in a quasi-static manner, they are all reversible and we return to the starting point after a cycle. In contrast to the case where *deterministic* data are erased as in the thermodynamically irreversible NAND operation, in the case we are discussing in this Box it is *random* data which are reversibly erased.

The cNOT gate copies the bit x if $y = 0$ and gives $\neg x$ if $y = 1$, and is the reversible equivalent of the COPY operation. It is reversible because there is a one-to-one correspondence between the initial state and the final state. The cNOT operation is a simple permutation of the basis vectors (see (5.4)). It can be shown that using single-bit gates

$$x \rightarrow 1 \oplus x \quad \text{or} \quad x \rightarrow \neg x$$

and the cNOT gate, it is possible to construct only linear functions if we limit ourselves to classical operations. If (x, y) and (x', y') are the initial and final bits, one can show that

$$\begin{pmatrix} x' \\ y' \end{pmatrix} = \begin{pmatrix} \alpha & \beta \\ \gamma & \delta \end{pmatrix} \begin{pmatrix} x \\ y \end{pmatrix} + \begin{pmatrix} \lambda \\ \mu \end{pmatrix}$$

where $\alpha, \ldots, \mu$ are numerical coefficients. It is necessary to introduce an additional gate, the *Toffoli gate*, which is a gate with three entrance and three exit bits, two of them control bits (x, y) and one a target bit z:

$$\text{Toffoli}: \quad (x, y, z) \rightarrow (x, y, z \oplus xy). \tag{5.3}$$

The nonlinearity of the gate is obvious from the xy factor. If $z = 1$, the Toffoli gate performs the NAND operation reversibly. The Toffoli gate can be used to reproduce reversibly all the classical logic circuits: the Toffoli gate is a universal gate for all the reversible operations of Boolean logic.

5.3 Quantum logic gates

The most general quantum evolution is a unitary transformation in the 2^n-dimensional Hilbert space of n qubits, $\mathcal{H}^{\otimes n}$. The most general quantum logic gate is a $2^n \times 2^n$ unitary matrix operating in $\mathcal{H}^{\otimes n}$. A theorem of linear algebra which we state without proof will allow us to limit ourselves to operations on one and two qubits.

Theorem Any unitary transformation on $\mathcal{H}^{\otimes n}$ can be decomposed into a product of cNOT gates and unitary transformations on one qubit.

As already explained, an operation on individual qubits cannot produce a general unitary transformation of $\mathcal{H}^{\otimes n}$, because such an operation has the form of a tensor product:

$$U = U^{(1)} \otimes U^{(2)} \otimes \cdots \otimes U^{(n)}.$$

It is necessary to perform nontrivial operations on at least two qubits to obtain a general unitary transformation. The above theorem guarantees that this is sufficient. This theorem is an existence theorem; in general, it is easier to construct the quantum logic gates for a given problem without using this theorem explicitly. It is useful to give the 4×4 matrix representation of the cNOT gate. In terms of qubits, the cNOT operation corresponds to the transformation

$$|00\rangle \rightarrow |00\rangle, \qquad |01\rangle \rightarrow |01\rangle, \qquad |10\rangle \rightarrow |11\rangle, \qquad |11\rangle \rightarrow |10\rangle.$$

In the basis $\{|00\rangle, |01\rangle, |10\rangle, |11\rangle\}$ the matrix representation then becomes

$$\text{cNOT} = \begin{pmatrix} 1 & 0 & 0 & 0 \\ 0 & 1 & 0 & 0 \\ 0 & 0 & 0 & 1 \\ 0 & 0 & 1 & 0 \end{pmatrix} = \begin{pmatrix} I & 0 \\ 0 & \sigma_x \end{pmatrix}. \tag{5.4}$$

In this form it is clear that cNOT cannot be a tensor product (Exercise 5.10.1). The generalization of the cNOT gate is the *control-U (cU) gate*, where the matrix σ_x is replaced by a 2×2 unitary matrix U:

$$\text{cU} = \begin{pmatrix} I & 0 \\ 0 & U \end{pmatrix}.$$

The cU gate leaves the target bit unchanged if $x = 0$ and modifies it as $|y\rangle \rightarrow U|y\rangle$ if $x = 1$. The cU gate can be constructed starting from the cNOT gate (Fig. 5.4). It is necessary to find three unitary operators A, B, and C such that

$$CBA = I, \qquad C\sigma_x B\sigma_x A = U.$$

In quantum physics, the Toffoli gate may be constructed from cU gates and cNOT gates (Fig. 5.4 with $U = \sqrt{\sigma_x}$) and the equation

$$\sqrt{\sigma_x} = \frac{1}{1+i} \begin{pmatrix} 1 & i \\ i & 1 \end{pmatrix},$$

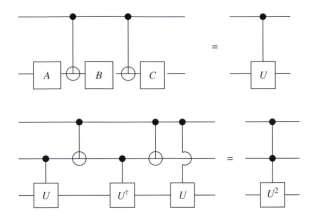

Figure 5.4 Construction of the cU gate and the Toffoli gate. The diagrams are read from left to right, and the products of operators act from right to left.

which is not possible in classical physics where the operation $\sqrt{\sigma_x}$ does not exist. In contrast to the classical case, it is not necessary to introduce the Toffoli gate explicitly to construct the ensemble of reversible logic circuits. We see from the results of Section 5.2 that if we have at our disposal a classical logic circuit allowing a function $f(x)$ to be calculated, then we can construct a quantum circuit using essentially the same number of gates. The justification of the circuits in Fig. 5.4 is left to Exercise 5.10.1.

Now that we know that there exists a quantum logic circuit which can evaluate a function $f(x)$, for example, by transposing a classical algorithm, we can state the basic ideas of *quantum parallelism*. We shall use two registers, an input register which stores x and an output register which stores the bits needed for $f(x)$. To simplify the discussion, we start with the case where the input register is a one-qubit register, as is the output register. We construct a transformation U_f (Fig. 5.5) which performs the operations

$$(x, y) \xrightarrow{U_f} (x, y \oplus f(x)).\tag{5.5}$$

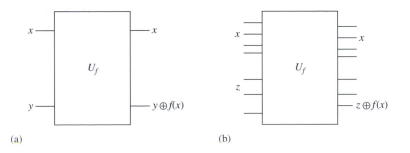

Figure 5.5 The construction of U_f: (a) 2 qubits, (b) $n + m$ qubits.

If the initial value is $y = 0$, we have simply

$$(x, 0) \xrightarrow{U_f} (x, f(x)).$$

One might ask why we do not simply perform the transformation $x \to f(x)$. The answer is that such a transformation cannot be unitary if the correspondence between x and $f(x)$ is not one-to-one, and so it is not suitable for a quantum algorithm. On the contrary, it is easy to convince ourselves that U_f is unitary, because its square is the identity:

$$(x, [y \oplus f(x)]) \xrightarrow{U_f} (x, [y \oplus f(x)] \oplus f(x)) = (x, y),$$

owing to $f(x) \oplus f(x) = 0$ for any $f(x)$. The operation U_f transforms one basis vector into another, and since $U_f^2 = I$ this correspondence can only be a simple permutation of the four basis vectors, and so it is a unitary transformation. In operator notation,

$$U_f |x \otimes 0\rangle = |x \otimes f(x)\rangle, \qquad U_f |x \otimes y\rangle = |x \otimes [y \oplus f(x)]\rangle. \qquad (5.6)$$

Let us apply to the state $|0\rangle_x$ a Hadamard gate H (not to be confused with the Hamiltonian $\hat{H}$):

$$\boxed{H = \frac{1}{\sqrt{2}} \begin{pmatrix} 1 & 1 \\ 1 & -1 \end{pmatrix}} \qquad (5.7)$$

or

$$H|0\rangle = \frac{1}{\sqrt{2}} (|0\rangle + |1\rangle), \qquad H|1\rangle = \frac{1}{\sqrt{2}} (|0\rangle - |1\rangle).$$

Now if the second qubit is in the initial state $|0\rangle$, the final state vector of the two qubits is the entangled state

$$|\Psi\rangle = U_f |H0 \otimes 0\rangle = U_f \frac{1}{\sqrt{2}} (|0 \otimes 0\rangle + |1 \otimes 0\rangle) = \frac{1}{\sqrt{2}} (|0 \otimes f(0)\rangle + |1 \otimes f(1)\rangle).$$

$$(5.8)$$

The state vector $|\Psi\rangle$ contains the information on $f(0)$ and $f(1)$ *simultaneously*, and the calculation of the vector $|\Psi\rangle$ *requires the same number of operations as that of* $U_f|0 \otimes 0\rangle$ *or* $U_f|1 \otimes 0\rangle$ *separately*: U_f is a unitary operator which does not depend on the state vector to which it is applied.

5.4 The Deutsch algorithm

Although $|\Psi\rangle$ in (5.8) contains information on $f(0)$ and $f(1)$ simultaneously, this does not give us any advantage over a classical computer if we wish to construct a table of values of $f(x)$ explicitly. However, it may happen that we need only information which does not require the construction of such a table. It is then possible that a quantum algorithm can exploit the information contained in $|\Psi\rangle$ to obtain the result using fewer operations than a classical algorithm. We shall explain how this works for the example of the Deutsch algorithm.

The Deutsch algorithm can be realized using the circuit of Fig. 5.6, with one-qubit input and output registers. The unknown function $f(x)$ takes the value 0 or 1 and we can ask the following question: do we have $f(0) = f(1)$ (a "constant" function) or $f(0) \neq f(1)$ (a "balanced" function)? If we were using a classical computer, we would have to calculate $f(0)$ and $f(1)$ and compare the two values. If we use a quantum computer, the question can be answered in a *single* operation. An equivalent problem is that of checking a coin: are the two sides different (a head *and* a tail) or are they the same (two heads or two tails)? The quantum computer allows us to make this comparison without looking at the two sides of the coin in succession.[4] This example is of course too elementary to be of any practical interest, but it gives the simplest illustration of quantum parallelism, and moreover it is a good warmup for the Grover algorithm of Section 5.6. The circuit of Fig. 5.6 gives the state $|\Psi\rangle$ at the entrance to the box U_f, the input register initially being in the state $|0\rangle$ and the output register in the state $|1\rangle$:

$$|\Psi\rangle = (H|0\rangle) \otimes (H|1\rangle) = \frac{1}{2}(|0\rangle + |1\rangle) \otimes (|0\rangle - |1\rangle) = \frac{1}{2}\left(\sum_{x=0}^{1}|x\rangle\right) \otimes (|0\rangle - |1\rangle).$$

(5.9)

We apply U_f (5.8) to this state with the following result:

1. if $f(x) = 0$, then $(|0\rangle - |1\rangle) \rightarrow (|0\rangle - |1\rangle)$,
2. if $f(x) = 1$, then $(|0\rangle - |1\rangle) = (|1\rangle - |0\rangle) \rightarrow -(|0\rangle - |1\rangle)$,

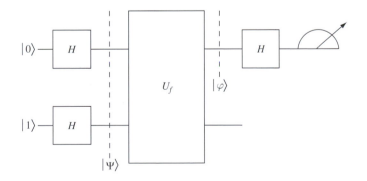

Figure 5.6 The Deutsch algorithm.

or, to summarize,

$$(|0\rangle - |1\rangle) \rightarrow (-1)^{f(x)} (|0\rangle - |1\rangle). \tag{5.10}$$

The state $U_f|\Psi\rangle$ is then the tensor product

$$U_f|\Psi\rangle = \frac{1}{2} \left(\sum_{x=0}^{1} (-1)^{f(x)} |x\rangle \right) \otimes (|0\rangle - |1\rangle). \tag{5.11}$$

The net result for the input register is

$$|x\rangle \xrightarrow{U_f} (-1)^{f(x)}|x\rangle. \tag{5.12}$$

In this particular case, the box U_f is called an *oracle*. Note that the input and output registers are unentangled after the oracle. The state of the qubit of the input register then is

$$|\varphi\rangle = \frac{1}{\sqrt{2}} \left((-1)^{f(0)}|0\rangle + (-1)^{f(1)}|1\rangle \right).$$

Before measuring the input register, we apply a Hadamard gate (see Fig. 5.6):

$$H|\varphi\rangle = \frac{1}{2} \left[(-1)^{f(0)} (|0\rangle + |1\rangle) + (-1)^{f(1)} (|0\rangle - |1\rangle) \right]$$
$$= \frac{1}{2} \left[(-1)^{f(0)} + (-1)^{f(1)} \right] |0\rangle + \frac{1}{2} \left[(-1)^{f(0)} - (-1)^{f(1)} \right] |1\rangle. \tag{5.13}$$

If measurement of the qubit gives $|0\rangle$, then $f(0) = f(1)$, i.e., the function is a "constant" one. If it gives $|1\rangle$, then $f(0) \neq f(1)$ and the function is a "balanced" one. The important point is that quantum parallelism has allowed us to bypass the explicit calculation of the function $f(x)$; the measurement of a single qubit contains the two possible results. The generalization to several qubits is left to Exercise 5.10.2.

5.5 Generalization to $n + m$ qubits

The above discussion can be generalized to an n-qubit input register and m-qubit output register, where m is the number of bits needed to write $f(x)$. We take as an example the case $n = 3$ for the input register. Using the notation $|x\rangle$, the number x is one of the eight numbers (in binary notation)

$$|000\rangle, \quad |001\rangle, \quad |010\rangle, \quad |011\rangle, \quad |100\rangle, \quad |101\rangle, \quad |110\rangle, \quad |111\rangle.$$

The special magic of a quantum computer is that it allows us to make linear combinations of the vectors of the computational basis using the operator H, which in the particular case $n = 3$ gives

$$|\Psi\rangle := H^{\otimes 3}|000\rangle = \frac{1}{\sqrt{8}} \sum_{x=0}^{7} |x\rangle,$$

where $H^{\otimes 3}$ denotes the tensor product of the three operators H. In general,

$$H^{\otimes n}|0^{\otimes n}\rangle = \frac{1}{2^{n/2}} \sum_{x=0}^{2^n-1} |x\rangle.$$

Here x is compact notation for the binary representation of the number x and the state vector of the computational basis is $|x\rangle = |x_{n-1} \cdots x_1 x_0\rangle$, where $x_{n-1}, \ldots, x_1, x_0$ take the value 0 or 1. The operator U_f is defined by generalizing the definition (5.6) as [5] (Fig. 5.5(b))

$$U_f|x \otimes z\rangle = |x \otimes [z \oplus f(x)]\rangle,$$

where $\oplus$ is mod 2 addition *without carry-over*, for example,

$$1101 \oplus 0111 = 1010.$$

We recall that

$$|x\rangle = |x_{n-1} \cdots x_1 x_0\rangle, \qquad |z\rangle = |z_{n-1} \cdots z_1 z_0\rangle$$

with $x_i, z_j = 0$ or 1. This assures that $U_f^2 = I$ and that U_f, which is a simple permutation of the 2^{n+m} basis vectors, is unitary. If we take $|0^{\otimes m}\rangle$ as the initial state of the output register, then

$$U_f|x \otimes 0^{\otimes m}\rangle = |x \otimes f(x)\rangle.$$

If finally we apply H to the input register in the state $|0^{\otimes n}\rangle$ before U_f, the state vector of the final state will be, by linearity,

$$|\Psi_{\text{fin}}\rangle = U_f|(H^{\otimes n}0^{\otimes n}) \otimes 0^{\otimes m}\rangle = \frac{1}{2^{n/2}} \sum_{x=0}^{2^n-1} |x \otimes f(x)\rangle. \qquad (5.14)$$

This state vector in principle contains the 2^n values of the function $f(x)$ (not necessarily all of them different). For example, if $n = 100$, it contains the $\sim 10^{30}$ values of $f(x)$: it is this exponential growth of states which leads to the miracle of quantum parallelism. A measurement can of course give only one of these values. As we have seen in the case of the Deutsch algorithm, it is nevertheless possible to extract useful information about the *relations* between the values of $f(x)$ for an ensemble of different values of x, of course at the price of losing the individual values. A classical computer, on the other hand, would have to evaluate $f(x)$ for all these values of x independently. In Section 5.7 we shall discuss this using the example of a quantum Fourier transform.

---

[5] Since later on we shall use y in a different context, here we denote the output register by z.

5.6 The Grover search algorithm

A quantum algorithm of more practical relevance than the Deutsch algorithm is the Grover search algorithm. This is an algorithm which performs a search for an entry in an *unstructured* data base, for example, a person's name in a telephone directory when the phone number is known. If N is the number of entries in the data base, a classical algorithm must on the average make $N/2$ attempts to find the name, as the only possibility is to check each entry one by one. The Grover algorithm allows the problem to be solved in $\sim \sqrt{N}$ operations.

The data base is stored using n qubits and we define the function $f(x)$, $x = \{0, 1, \ldots, 2^n - 1\}$, such that $f(x) = 0$ if $x \neq y$ and $f(x) = 1$ if $x = y$ is a solution: $f(x) = \delta_{xy}$. To simplify the argument, we assume that the value of y is unique. We define an operator O, the oracle, whose action in the computational basis is (see (5.12))

$$O|x\rangle = (-1)^{f(x)}|x\rangle. \tag{5.15}$$

As in the case of the Deutsch algorithm (see (5.11)), the auxiliary qubit (lowest qubit in Fig. 5.8) is unentangled with the other qubits after the oracle. The Grover operator G is defined as

$$G = H^{\otimes n} X H^{\otimes n} O = H^{\otimes n} (2|0\rangle\langle 0| - I) H^{\otimes n} O, \tag{5.16}$$

where

$$X|x\rangle = -(-1)^{\delta_{x0}}|x\rangle = (2|0\rangle\langle 0| - I)|x\rangle.$$

To simplify the notation, we shall introduce the vector $|\Psi\rangle$ already used in Section 5.5:

$$|\Psi\rangle = H^{\otimes n}|0^{\otimes n}\rangle = \frac{1}{2^{n/2}} \sum_{x=0}^{2^n - 1} |x\rangle. \tag{5.17}$$

Taking into account $H^2 = I$, we find

$$H^{\otimes n} (2|0\rangle\langle 0| - I) H^{\otimes n} = 2H^{\otimes n}|0\rangle\langle 0|H^{\otimes n} - I = 2|\Psi\rangle\langle\Psi| - I$$

and so

$$G = (2|\Psi\rangle\langle\Psi| - I) O. \tag{5.18}$$

This construction can be used to draw the quantum logic circuit corresponding to G (see Fig. 5.8(b)).

The operator G can be interpreted as a rotation in a two-dimensional plane. Let $|\alpha\rangle$ be the normalized vector

$$|\alpha\rangle = \frac{1}{\sqrt{N-1}} \sum_{x \neq y} |x\rangle \tag{5.19}$$

$(N = 2^n)$, which can be used to write $|\Psi\rangle$ as

$$|\Psi\rangle = \sqrt{1 - \frac{1}{N}}\, |\alpha\rangle + \sqrt{\frac{1}{N}}\, |y\rangle. \tag{5.20}$$

We rewrite this equation as

$$|\Psi\rangle = \cos\frac{\theta}{2}\, |\alpha\rangle + \sin\frac{\theta}{2}\, |y\rangle, \tag{5.21}$$

where the angle θ is given by

$$\cos\frac{\theta}{2} = \sqrt{1 - \frac{1}{N}}.$$

According to (5.15), the action of the oracle on $|\Psi\rangle$ is

$$O(\lambda|\alpha\rangle + \mu|y\rangle) = \lambda|\alpha\rangle - \mu|y\rangle.$$

This is a reflection with respect to the direction of $|\alpha\rangle$ in the plane Π subtended by $|\alpha\rangle$ and $|y\rangle$ (Fig. 5.7). Moreover, $(2|\Psi\rangle\langle\Psi| - I)$ performs a reflection in Π with respect to the direction of $|\Psi\rangle$: if $\langle\Psi|\Phi\rangle = 0$,

$$(2|\Psi\rangle\langle\Psi| - I)(\lambda|\Psi\rangle + \mu|\Phi\rangle) = \lambda|\Psi\rangle - \mu|\Phi\rangle.$$

However, the product of two reflections is a rotation, and Fig. 5.7 shows that the angle taking us from $|\alpha\rangle$ to $G|\Psi\rangle$ is $3\theta/2$:

$$G|\Psi\rangle = \cos\frac{3\theta}{2}\, |\alpha\rangle + \sin\frac{3\theta}{2}\, |y\rangle. \tag{5.22}$$

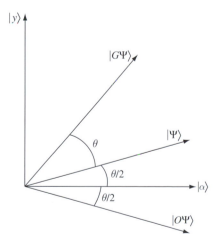

Figure 5.7 Schematic depiction of the rotations and reflections of the Grover algorithm.

The angle between $|\Psi\rangle$ and $G|\Psi\rangle$ is θ, and the angle between $|\alpha\rangle$ and $G|\Psi\rangle$ is $3\theta/2$; $G|\Psi\rangle$ is deduced from $|\Psi\rangle$ by rotation by an angle θ. Moreover, $G^2|\Psi\rangle$ is deduced from $G|\Psi\rangle$ by rotation by an angle θ. After k iterations of G, $G^k|\Psi\rangle$ is always in Π and is deduced from $|\alpha\rangle$ by rotation by an angle $(2k+1)\theta/2$:

$$G^k|\Psi\rangle = \cos\frac{(2k+1)\theta}{2}\,|\alpha\rangle + \sin\frac{(2k+1)\theta}{2}\,|y\rangle. \qquad (5.23)$$

The effect of successive rotations is to make $G^k|\Psi\rangle$ come closer and closer to $|y\rangle$. The optimal value $k = k_0$ of k is determined using the following argument. We wish to have

$$0 = \cos\frac{(2k+1)\theta}{2} = \cos k\theta \cos\frac{\theta}{2} - \sin k\theta \sin\frac{\theta}{2}$$

$$= \sqrt{1 - \frac{1}{N}}\,\cos k\theta - \sqrt{\frac{1}{N}}\,\sin k\theta.$$

We then find that $\tan k\theta = \sqrt{N-1}$, or $\cos k\theta = 1/\sqrt{N}$, and so

$$k_0 = \left[\frac{1}{\theta}\cos^{-1}\sqrt{\frac{1}{N}}\right] + 1,$$

where $[x]$ is the integer part of x. For $N \gg 1$ we have, comparing (5.20) and (5.21), $\theta \simeq 2/\sqrt{N}$ or

$$k_0 \simeq \frac{\sqrt{N}}{2}\cos^{-1}\sqrt{\frac{1}{N}} \simeq \frac{\pi\sqrt{N}}{4}. \qquad (5.24)$$

It is therefore sufficient to apply the oracle $\sim\sqrt{N}$ times in order to have a very good chance of obtaining the result. To estimate the probability of this, we

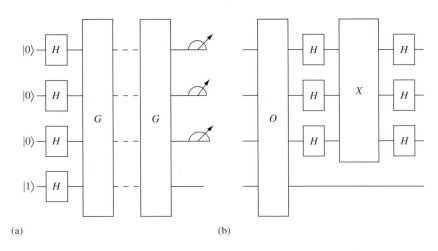

(a)   (b)

Figure 5.8 (a) Logic circuits of the Grover algorithm for $n = 3$. (b) The circuits of G. The action of the oracle O is $O|x\rangle = (-1)^{f(x)}|x\rangle$ and that of the box X is $X|x\rangle = -(-1)^{\delta_{x0}}|x\rangle$.

note that according to Fig. 5.7 the angle between $G^{k_0}|\Psi\rangle$ and $|y\rangle$ is less than $\theta/2$. The probability of error is therefore less than $\mathcal{O}(1/N)$. It can be shown that the Grover algorithm is optimal: it is not possible to find a faster algorithm than Grover's. If we count the quantum logic gates, the total number of operations of the Grover algorithm is actually $\simeq \sqrt{N} \ln N$. The circuit corresponding to the Grover algorithm is shown schematically in Fig. 5.8.

5.7 The quantum Fourier transform

The last algorithm we shall describe is that of Shor. As a preliminary, let us construct a quantum logic circuit for the Fourier transform. Let an integer x, $0 \le x \le 2^n - 1$, be written using n bits

$$x = 0, 1, \ldots, 2^n - 1,$$

and let $|x\rangle$ be a vector of the computational basis

$$|x\rangle = |x_{n-1} \cdots x_1 x_0\rangle, \qquad x_i = 0 \text{ or } 1.$$

We define a unitary transformation[6] U_{FT} whose matrix elements in the computational basis are

$$\langle y|U_{\mathrm{FT}} x\rangle = (U_{\mathrm{FT}})_{yx} = \frac{1}{2^{n/2}} e^{2i\pi xy/2^n}. \tag{5.25}$$

The transformation U_{FT} is physically realized in the box U_{FT} of Fig. 5.9(a), and, as we shall soon see, a possible circuit is that given in Fig. 5.9(b). If $|\Psi\rangle$ is a normalized linear combination of the vectors $|x\rangle$,

$$|\Psi\rangle = \sum_{x=0}^{2^n-1} f(x)|x\rangle, \qquad \sum_{x=0}^{2^n-1} |f(x)|^2 = 1, \tag{5.26}$$

where $f(x) = \langle x|\Psi\rangle$, then the amplitude for finding at the exit from the box U_{FT} a state $|y\rangle$ of the computational basis (note that $|y\rangle$ denotes a state of the input register) is, from (2.17) setting $|\Phi\rangle = U_{\mathrm{FT}}|\Psi\rangle$,

$$a(\Phi \to y) = \langle y|\Phi\rangle = \sum_{x=0}^{2^n-1} \langle y|U_{\mathrm{FT}} x\rangle\langle x|\Psi\rangle$$

$$= \frac{1}{2^{n/2}} \sum_{x=0}^{2^n-1} e^{2i\pi xy/2^n} f(x) = \tilde{f}(y), \tag{5.27}$$

---

[6] In fact,

$$\sum_{y=0}^{2^n-1} (U_{\mathrm{FT}}^\dagger)_{x'y}(U_{\mathrm{FT}})_{yx} = \sum_{y=0}^{2^n-1} (U_{\mathrm{FT}}^*)_{yx'}(U_{\mathrm{FT}})_{yx} = \frac{1}{2^{n/2}} \sum_{y=0}^{2^n-1} e^{2i\pi(x-x')y/2^n} = \delta_{x'x}.$$

The result is obtained upon noticing that the sum over y is a geometric series.

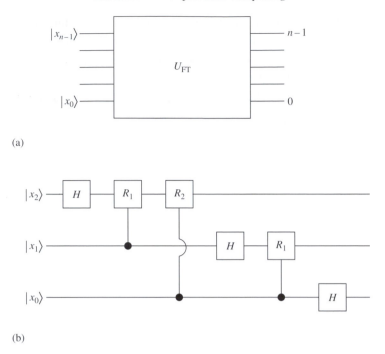

(a)

(b)

Figure 5.9 (a) The box U_{FT}. (b) A circuit constructing U_{FT} in the case $n = 3$.

where we have used the completeness relation $\sum_x |x\rangle\langle x| = I$ (Box 2.1). The probability amplitude $a(\Phi \to y)$ is just the discrete (or lattice) Fourier transform $\tilde{f}(y)$ of $f(x)$.

For constructing the box U_{FT} it is convenient to write $U_{\mathrm{FT}}|x\rangle$ as

$$U_{\mathrm{FT}}|x\rangle = \sum_{y=0}^{2^n-1} |y\rangle\langle y|U_{\mathrm{FT}}\,x\rangle = \frac{1}{2^{n/2}} \sum_{y=0}^{2^n-1} e^{2i\pi xy/2^n} |y\rangle. \tag{5.28}$$

We shall transform (5.28) so as to write it as a manifestly nonentangled state using a standard technique of fast Fourier transforms. Let

$$x = x_0 + 2x_1 + \cdots + 2^{n-1}x_{n-1},$$
$$y = y_0 + 2y_1 + \cdots + 2^{n-1}y_{n-1}, \tag{5.29}$$

be the binary decompositions of x and y and let us examine the factor $xy/2^n$ in the exponential of (5.28) in the case $n = 3$, $N = 2^3 = 8$. Using the fact that $\exp(2i\pi p) = 1$ for integer p, we can replace the product $xy/8$ in the exponential of (5.28) by

$$\frac{xy}{8} \to y_0\left(\frac{x_2}{2} + \frac{x_1}{4} + \frac{x_0}{8}\right) + y_1\left(\frac{x_1}{2} + \frac{x_0}{4}\right) + y_2\frac{x_0}{2}$$
$$= y_0.x_2x_1x_0 + y_1.x_1x_0 + y_2.x_0,$$

where we have introduced the notation (the binary representation of a number less than one)

$$.x_p x_{p-1} \cdots x_1 x_0 = \frac{x_p}{2} + \frac{x_{p-1}}{2^2} + \cdots + \frac{x_0}{2^p}. \tag{5.30}$$

Coming back to the general case, we see that it is now possible to factorize the sum over y into sums over $y_0, \ldots, y_{n-1}$, $y_i = 0$ or 1, $|y\rangle = |y_{n-1} \cdots y_1 y_0\rangle$:

$$U_{FT}|x\rangle = \frac{1}{2^{n/2}} \sum_{y_0,\ldots,y_{n-1}} e^{2i\pi y_{n-1}.x_0} \cdots e^{2i\pi y_0.x_{n-1}x_0} \cdots |y_{n-1} \cdots y_0\rangle$$

$$= \frac{1}{2^{n/2}} \left(\sum_{y_{n-1}} e^{2i\pi y_{n-1}.x_0} |y_{n-1}\rangle \right) \cdots \left(\sum_{y_1} e^{2i\pi y_1.x_{n-2}\cdots x_0} |y_1\rangle \right)$$

$$\times \left(\sum_{y_0} e^{2i\pi y_0.x_{n-1}\cdots x_0} |y_0\rangle \right),$$

or, expanding each quantity in parentheses,

$$U_{FT}|x\rangle = \frac{1}{2^{n/2}} \left(|0\rangle_{n-1} + e^{2i\pi.x_0}|1\rangle_{n-1} \right) \cdots \left(|0\rangle_1 + e^{2i\pi.x_{n-2}\cdots x_0}|1\rangle_1 \right)$$

$$\times \left(|0\rangle_0 + e^{2i\pi.x_{n-1}\cdots x_0}|1\rangle_0 \right), \tag{5.31}$$

which manifestly has the form of a tensor product. Let us give an example for $n = 2$:

$$U_{FT}|x\rangle \equiv U_{FT}|x_1 x_0\rangle = \frac{1}{4} \left(|00\rangle + e^{2i\pi.x_1 x_0}|01\rangle + e^{2i\pi.x_0}|10\rangle + e^{2i\pi(.x_1 x_0 + .x_0)}|11\rangle \right)$$

$$= \frac{1}{4} \left(|0\rangle_1 + e^{2i\pi.x_0}|1\rangle_1 \right) \left(|0\rangle_0 + e^{2i\pi.x_1 x_0}|1\rangle_0 \right).$$

A possible logic circuit for performing this Fourier transform is shown in Fig. 5.9(b). The gate cR_d is defined by the operator R_d:

$$R_d = \begin{pmatrix} 1 & 0 \\ 0 & e^{i\pi/2^d} \end{pmatrix}. \tag{5.32}$$

Let us study the circuit of Fig. 5.9(b). The action of the gate H is

$$H|0\rangle_2 = \frac{1}{\sqrt{2}} \left(|0\rangle_2 + |1\rangle_2 \right), \qquad H|1\rangle_2 = \frac{1}{\sqrt{2}} \left(|0\rangle_2 - |1\rangle_2 \right),$$

and so the action on the first bit $|x_2\rangle$ can be written as

$$H|x_2\rangle = \frac{1}{\sqrt{2}} \left(|0\rangle_2 + e^{2i\pi.x_2}|1\rangle_2 \right). \tag{5.33}$$

We use $c_i R_d^j$ to denote the action on the bit j of R_d controlled by the bit i. Then

$$x_1 = 0: \qquad (c_1 R_1^2) H |x_2\rangle = \frac{1}{\sqrt{2}} \left(|0\rangle_2 + e^{2i\pi . x_2} |1\rangle_2 \right),$$

$$x_1 = 1: \qquad (c_1 R_1^2) H |x_2\rangle = \frac{1}{\sqrt{2}} \left(|0\rangle_2 + e^{2i\pi . x_2} e^{i\pi/2} |1\rangle_2 \right),$$

which can be written as

$$(c_1 R_1^2) H |x_2\rangle = \frac{1}{\sqrt{2}} \left(|0\rangle_2 + e^{2i\pi . x_2 x_1} |1\rangle_2 \right). \tag{5.34}$$

It is clear that the procedure is followed by

$$(c_0 R_2^2)(c_1 R_1^2) H |x_2\rangle = \frac{1}{\sqrt{2}} \left(|0\rangle_2 + e^{2i\pi . x_2 x_1 x_0} |1\rangle_2 \right), \tag{5.35}$$

and after applying all the gates in Fig. 5.9(b) we obtain the state

$$|\Psi'\rangle = \frac{1}{\sqrt{8}} \left(|0\rangle_0 + e^{2i\pi . x_0} |1\rangle_0 \right) \left(|0\rangle_1 + e^{2i\pi . x_1 x_0} |1\rangle_1 \right) \left(|0\rangle_2 + e^{2i\pi . x_2 x_1 x_0} |1\rangle_2 \right).$$

The qubits are in the wrong order, and one may use SWAP gates to put them in the right one. However, one can also write by convention the computational basis in the order

$$|x\rangle = |x_0 x_1 \cdots x_{n-1}\rangle$$

that is, the first digit is x_0 and the last digit x_{n-1}, meaning that the number is read from right to left. Then one can avoid the SWAP gates altogether. The number of gates needed decomposes into n gates H and

$$n + (n-1) + \cdots + 1 \simeq \frac{1}{2} n^2$$

conditional gates cR_d, or $\mathcal{O}(n^2)$ gates.

5.8 The period of a function

The Shor factorization algorithm is based on the possibility of "rapidly," that is, in a time which is a polynomial in n, finding the period of a function $f(x)$, which in the Shor case is the function $b^x \bmod N$. Let us assume that we have a function $f(x)$ of period r, $f(x) = f(x+r)$, with

$$x = 0, 1, \ldots, 2^n - 1. \tag{5.36}$$

For the algorithm to be successful we must have, as we shall see, $2^n > N^2$. A classical algorithm uses $\mathcal{O}(N)$ elementary operations (the function $b^x \bmod N$ appears to be random noise over a period and does not give any key to what the period is), but the quantum algorithm we shall describe below uses only $\mathcal{O}(n^3)$ elementary operations. The variable x is stored in a register $|x\rangle$ and the function $f(x)$ in a register $|z\rangle$ corresponding to m qubits. We start from an initial state of $n+m$ qubits:

$$|\Phi\rangle = \frac{1}{2^{n/2}} \left(\sum_{x=0}^{2^n-1} |x\rangle \right) \otimes |0\cdots0\rangle. \tag{5.37}$$

We then use the box U_f which calculates the function $f(x)$:

$$|\Psi_f\rangle = U_f|\Phi\rangle = \frac{1}{2^{n/2}} \sum_{x=0}^{2^n-1} |x \otimes f(x)\rangle. \tag{5.38}$$

This requires $\mathcal{O}(n)$ operations. If we measure the output register and find the result f_0, the state vector of the input register after this measurement is given by the state vector collapse (see Section 2.4)

$$|\Psi_0\rangle = \frac{1}{\mathcal{N}} \sum_{x;\, f(x)=f_0} |x\rangle, \tag{5.39}$$

where the sum runs over the values of x such that $f(x) = f_0$, and $\mathcal{N}$ is a normalization factor. We shall assume that $f(x+s) = f(x)$ implies that $s = pr$, p integer, in other words, the function $f(x)$ never takes the same value twice in a period, which is the case for the function $b^x \bmod N$. The normalized vector $|\Psi\rangle$ of the input register, with $f(x_0) = f_0$ and x_0 being the smallest value of x such that $f(x_0) = f_0$, then is

$$|\Psi_0\rangle = \frac{1}{\sqrt{K}} \sum_{k=0}^{K-1} |x_0 + kr\rangle, \tag{5.40}$$

where [7] $K \simeq 2^n/r$. *In reality, it is not necessary to measure the output register* (Box 5.2). At the exit from the box U_f of Fig. 5.10, the qubits of the input register are entangled with the qubits of the output register (see (5.38)), and if only the qubits of the input register are observed, it is necessary to take the trace over the output register in order to obtain the state operator of the qubits of the input register; the physical state of the qubits of the input register will in general be described by a state operator, and not by a vector of $\mathcal{H}^{\otimes n}$. Stated differently, the physical state of the input register is an *incoherent* superposition of the vectors $|\Psi_i\rangle$:

$$|\Psi_i\rangle = \frac{1}{\sqrt{K_i}} \sum_{k=0}^{K_i-1} |x_i + kr\rangle, \tag{5.41}$$

where $f(x_i) = f_i$ and x_i is the smallest value of x such that $f(x_i) = f_i$. Since the rest of the argument does not depend on x_i, we can just as well avoid measuring the output register. In other words, it is completely unnecessary to resort to state vector collapse.

[7] More precisely, $K = [2^n/r]$ or $K = [2^n/r]+1$, where $[z]$ denotes the integer part of z.

Box 5.2: What measurements are needed?

Formally, the *total* state operator (of the input and output registers) ρ_{tot} is, according to (5.38),

$$\rho_{\text{tot}} = |\Psi_f\rangle\langle\Psi_f| = \frac{1}{2^n}\sum_{x,z}|x\otimes f(x)\rangle\langle z\otimes f(z)|.$$

The state operator of the input register is obtained by taking the partial trace over the output register (see (4.14)):

$$\rho_{\text{in}} = \text{Tr}_{\text{out}}\rho_{\text{tot}} = \frac{1}{2^n}\sum_{x,z}|x\rangle\langle z|\,\langle f(z)|f(x)\rangle.$$

Let us suppose that the function $f(x)$ takes the value f_0 N_0 times, and the value f_1 N_1 times, with $N_0 + N_1 = 2^n$. Then

$$\rho_{\text{in}} = \frac{1}{2^n}\left(\sum_{x,z;f(x)=f(z)=f_0}|x\rangle\langle z| + \sum_{x,z;f(x)=f(z)=f_1}|x\rangle\langle z|\right),$$

because $\langle f(x)|f(z)\rangle = 1$ if $f(x) = f(z)$ and $\langle f(x)|f(z)\rangle = 0$ if $f(x) \neq f(z)$. This corresponds to an incoherent superposition of normalized vectors

$$|\Psi_0\rangle = \frac{1}{\sqrt{N_0}}\sum_{x;f(x)=f_0}|x\rangle, \quad |\Psi_1\rangle = \frac{1}{\sqrt{N_1}}\sum_{x;f(x)=f_1}|x\rangle$$

with the probabilities $\mathsf{p}_0 = N_0/2^n$ and $\mathsf{p}_1 = N_1/2^n$. In the case of our periodic function, the reduced state operator of the input register is

$$\rho_{\text{in}} = \frac{1}{2^n}\sum_{i=0}^{r-1}\sum_{k_i,k_j=0}^{K_i-1}|x_i+k_i r\rangle\langle x_i+k_j r|.$$

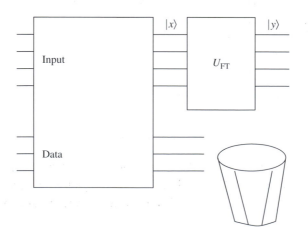

Figure 5.10 Schematic depiction of calculation of the period. The qubits of the output register are discarded.

The state vector (5.40) corresponds in (5.26) to the choice $f(x) = 1/\sqrt{K}$ if x has the form $x_0 + kr$ and $f(x) = 0$ otherwise. According to (5.27), the amplitude $a(\Phi_0 \to y)$, where $|\Phi_0\rangle = U_{FT}|\Psi_0\rangle$, then is

$$a(\Phi_0 \to y) = \frac{1}{2^{n/2}} \frac{1}{\sqrt{K}} \sum_{k=0}^{K-1} e^{2i\pi y(x_0+kr)/2^n}, \tag{5.42}$$

and the probability of measuring the value y (that is, of finding the state $|y_{n-1}\cdots y_1 y_0\rangle$ of the computational basis at the exit from the box U_{FT}) is

$$p(y) = |a(\Phi_0 \to y)|^2 = \frac{1}{2^n K} \left| \sum_{k=0}^{K-1} e^{2i\pi kry/2^n} \right|^2. \tag{5.43}$$

We observe that $p(y)$ is independent of x_0, and we could have started from any of the vectors $|\Psi_i\rangle$ of (5.41). We next use the geometric series [8]

$$\sum_{k=0}^{K-1} e^{2i\pi ykr/2^n} = \frac{1 - e^{2i\pi yKr/2^n}}{1 - e^{2i\pi yr/2^n}} = e^{i\pi(K-1)r/2^n} \frac{\sin(\pi yKr/2^n)}{\sin(\pi yr/2^n)}.$$

In the exceptional case that $2^n/r$ is an integer, and therefore $2^n/r = K$, we would find (recall that y is an integer)

$$p(y) = \frac{1}{2^n K} \frac{\sin^2(\pi y)}{\sin^2(\pi y/K)} = \frac{1}{r} \quad \text{if} \quad y = jK$$

$$= 0 \quad \text{otherwise,}$$

where j is an integer. We then find that $j/r = y/2^n$, which gives j and r if we are lucky and j/r happens to be an irreducible fraction. We cast $y/2^n$ into its irreducible form j_0/r_0, and then $j = j_0$, $r = r_0$. It is clear that the desired result has been obtained from constructive interference. In the general case we can write, always with integer j (but noninteger $2^n/r$!),

$$y_j = j\frac{2^n}{r} + \delta_j, \tag{5.44}$$

which gives the probability $p(y_j)$:

$$p(y_j) = \frac{1}{2^n K} \frac{\sin^2(\pi \delta_j Kr/2^n)}{\sin^2(\pi \delta_j r/2^n)}. \tag{5.45}$$

In general, the function $p(y)$ has sharp maxima when the value of y is close to $j2^n/r$. Using the bounds on $\sin x$,

$$\frac{2}{\pi}x \le \sin x \le x, \qquad 0 \le x \le \frac{\pi}{2},$$

---

[8] The problem is reminiscent of that of diffraction, for example, neutron diffraction by a crystal. If a is the distance between two sites ($a = 1$ in the text), the lattice cell is ra. The (quasi-) wavevector q can take the values $q = 2\pi p/2^n a$, $p = 0, 1, \ldots, 2^n - 1$ (p and $p' = p + 2^n$ are equivalent). Diffraction peaks are produced when q is an integer multiple of the reciprocal lattice cell $2\pi/ra$, or $q = j2\pi/ra$, $j = 0, 1, \ldots, r-1$.

we can show that the probability of reaching one of the values (5.44) if we require $|\delta_j| < 1/2$ is at least $4/\pi^2$:

$$p(y_j) \geq \frac{4}{\pi^2} \frac{K}{2^n} \simeq \frac{4}{\pi^2} \frac{1}{r}.$$

Since $0 \leq j \leq r - 1$ and $r \gg 1$, there is at least a 40% ($4/\pi^2 \simeq 0.406$) chance of finding one value of y_j close to $j2^n/r$. More precisely (see Exercise 5.10.3 for a specific example),

$$\left| y_j - j\frac{2^n}{r} \right| \leq \frac{1}{2}.$$

Since n and y_j are known (y_j is an integer, $0 \leq y_j \leq 2^n - 1$, the result of measuring the input register), we therefore have an estimate of the fraction j/r. Let us now show that measurement of y_j allows j and r to be determined (always with at least 40% probability). We let y_j vary by one unit, which gives

$$\left| (y_j \pm 1) - j\frac{2^n}{r} \right| \geq \frac{1}{2},$$

in contradiction with the preceding equation, and the (integer) value of y_j is determined by the condition $|\delta_j| < 1/2$. Owing to our choice $2^n > N^2$, which implies that $2^n > r^2$, we have obtained an estimate of j/r which differs from the exact value by less than $1/2r^2$:

$$\left| \frac{y_j}{2^n} - \frac{j}{r} \right| \leq \frac{1}{2^{n+1}}. \tag{5.46}$$

Since $r < N$, and since two fractions of denominator $\geq r$ must differ by at least $1/r^2$ unless they are identical,[9] we will then have a unique value of the fraction j/r. The value of j/r can be determined from the known value of $y_j/2^n$ by expansion in continued fractions, which gives the value of j/r as an irreducible fraction j_0/r_0. If we are lucky and j and r have no common factor, we will immediately obtain the value $r = r_0$ for r. The probability that two large numbers have no common factor is at least [10] 60%, and, with a probability $\sim 0.4 \times 0.6$ or once in four times, the method will directly give the period $r = r_0$, as can be

---

[9] Because

$$\left| \frac{n}{m} - \frac{p}{q} \right| \geq \frac{1}{mq}$$

unless the two fractions are identical.

[10] There is one chance in two that a number is divisible by 2, one in three that it is divisible by 3,..., one in p that it is divisible by p, ... The probability for two large numbers to be divisible simultaneously by p is $1/p^2$, and the probability that they have no common factor is

$$\prod_{p=2}^{\infty} \left(1 - \frac{1}{p^2} \right) = \frac{1}{\zeta(2)} = \frac{6}{\pi^2}.$$

verified using a classical computer by comparing $f(x)$ and $f(x+r_0)$. If $f(x) \neq f(x+r_0)$, we can try out the first multiples of $r_0 : 2r_0, 3r_0, \ldots$, and if these trials give nothing, this indicates that the value of y_j was probably outside the interval $|\delta_j| \leq 1/2$. It is then necessary to repeat the entire procedure and measure another value for y_j. This procedure takes $\mathcal{O}(n^3)$ elementary operations, $\mathcal{O}(n^2)$ for the Fourier transform and $\mathcal{O}(n)$ for the calculation of b^x.

Determination of the period r is sufficient to crack the RSA code. In effect (Box 2.2), Eve has at her disposal the message encoded by Alice, b, and the numbers N and c, which are available publicly. She calculates d' as $cd' \equiv 1 \bmod r$ and then $b^{d'} \bmod N$:

$$b^{d'} = a^{cd'} = a^{1+mr} = a(a^r)^m \equiv a \bmod N,$$

because $a^r \equiv 1 \bmod N$ (Box 5.3), and Eve then recovers the original message a.

Box 5.3: The mathematics of RSA encryption

Let N be an integer and G_N be the set of integers $<N$ which have no common factor with N. If $a \in G_N$, then a and N have no common factor. G_N is closed under mod N multiplication, because if $a, b \in G_N$, then $ab \bmod N \in G_N$. In fact, the product ab can be written as

$$ab = x + qN, \qquad ab \equiv x \bmod N,$$

where x cannot have a common factor with N. If x had a common factor s with N, it would be possible to write

$$ab = s(x' + qN/s)$$

and ab would then have s as a factor, which is impossible. Furthermore, if $a, b, c \in G_N$ and $ab \equiv ac \bmod N$, then

$$a(b-c) = pN.$$

Since a has no common factor with N, $(b-c)$ must be a multiple of N, and because $b, c < N$, this means that $b = c$. As a result, if $b \neq c$, $ab \bmod N$ and $ac \bmod N$ will be different and the multiplication of the elements of G_N by a is a simple permutation of these elements. Since $1 \in G_N$, it follows that a possesses an inverse d in G_N, $ad \equiv 1 \bmod N$, and G_N is therefore a group. The order k of an element a of G_N is the smallest integer k such that $a^k \equiv 1 \bmod N$; the integer k is a divisor [11] of the order (the number of elements) of G_N. If N is a prime, the order of G_N is $(N-1)$, and then $\forall a < N$, and therefore also for any a not divisible by N

$$a^{N-1} \equiv 1 \bmod N,$$

---

[11] The order of a subgroup is a divisor of the order of the group: $1, a, a^2, \ldots, a^{k-1}$ form a subgroup of G_N.

because $(N-1)$ is a multiple of k. Let us take two primes p and q and an integer a which is divisible by neither p nor q; a^{q-1} is not divisible by p, so

$$[a^{q-1}]^{(p-1)} \equiv 1 \bmod p$$

and similarly a^{p-1} is not divisible by q

$$[a^{p-1}]^{(q-1)} \equiv 1 \bmod q,$$

that is,

$$a^{(p-1)(q-1)} = 1 + mp, \qquad a^{(p-1)(q-1)} = 1 + nq,$$

which implies that $mp = nq$ and so

$$a^{(p-1)(q-1)} = 1 + kpq, \qquad a^{(p-1)(q-1)} \equiv 1 \bmod pq.$$

We then deduce that

$$a^{1+s(p-1)(q-1)} \equiv a \bmod pq.$$

This last relation is valid whatever a, that is, even if a can be divided by p (or q), as the reader can easily check.

If c does not have a common factor with $(p-1)(q-1)$, then it has an inverse d in $G_{(p-1)(q-1)}$:

$$cd \equiv 1 \bmod (p-1)(q-1), \qquad cd = 1 + s(p-1)(q-1),$$

and if $b \equiv a^c \bmod pq$, then

$$b^d = a^{cd} \equiv a \bmod pq,$$

which gives the formula on which RSA encryption is based (Box 2.3).
The subgroups generated by a and b are the same because $b = a^c$. Let r be the order of this subgroup. We may assume that the integers $a, b \in G_{pq}$;[12] then r must be a divisor of $(p-1)(q-1)$, but since c has no common factor with $(p-1)(q-1)$ it cannot have a common factor with r. As a result, $c \in G_r$ and there exists a d' such that

$$cd' \equiv 1 \bmod r.$$

Since Eve knows r, she can calculate d' using $cd' \equiv 1 \bmod r$ and then $b^{d'}$ mod N:

$$b^{d'} = a^{cd'} = a^{1+mr} = a(a^r)^m \equiv a \bmod N,$$

because $a^r \equiv 1 \bmod N$, and so Eve recovers the original message a.

---

[12] If this is not the case so that, for example, b and N have a common divisor (which is unlikely as b and N are very large numbers), Eve could first compute $\gcd(b, N)$. Then she would have directly factored N.

If we wish in addition to factorize N we must write

(i)

$$a^r - 1 = \left(a^{r/2} - 1\right)\left(a^{r/2} + 1\right) \equiv 0 \bmod N,$$

(ii)

$$a^{r/2} \not\equiv \pm 1 \bmod N.$$

If we are lucky, that is, if the two factors in (i) are integers and (ii) is satisfied, then the product of integers

$$\left(a^{r/2} - 1\right)\left(a^{r/2} + 1\right)$$

is divisible by $N = pq$. It is therefore necessary that p divides $\left(a^{r/2} - 1\right)$ and q divides $\left(a^{r/2} + 1\right)$ or vice versa. The values of p and q are obtained by seeking the greatest common divisors (gcds):

$$p = \gcd\left(N, a^{r/2} - 1\right), \qquad q = \gcd\left(N, a^{r/2} + 1\right).$$

If we are unlucky, we must start over, but the probability of success is greater than 50%. It is worth noting that the algorithm we have just described is a *probabilistic algorithm*: it does not work every time, but it has a good chance of working, and we can be sure that it will work after a small number of tries.

5.9 Classical algorithms and quantum algorithms

The theory of quantum algorithms raises doubts about some of the statements of the theory of classical algorithms when the subject of *algorithmic complexity* is considered, i.e., when we ask what resources are needed to perform a calculation. A general idea is that certain problems can be solved in a number of steps $\mathcal{N}$ which is a polynomial in the number of bits n that measures the size of the problem. For example, if we wish to multiply two numbers n binary digits long, the number of instructions needed is a polynomial in n. A much less trivial example is that of finding primes: how many steps are needed to show that a number is a prime? In 2002 it was proved that this problem is polynomial. However, experience suggests that other problems require a number of computational steps which grows more rapidly than any power of n for $n \gg 1$, for example, as $\exp n$, $\exp(n^{1/3})$, or $n^{\ln n}$. Such problems are often, somewhat inaccurately, termed "exponential."

Turing defined a class of machines, now known as *Turing machines*, which made it possible to study the concept of the complexity of a computational algorithm. He showed that there exist machines, called universal machines (and he also proposed a design for one), which are capable of simulating any other Turing machine. Since then it has been discovered that all computational models proposed for executing programs can be simulated by a Turing machine in a computational time which is a polynomial in the computational time of the simulated machine. This result suggested the following generalization. All machines are equivalent as regards the computational time (or the number of computational steps), up to a polynomial. If this idea is correct, the exponential or polynomial nature of a problem is preserved in going from one computational model to another, which leads to the idea of precisely defining the algorithmic complexity of a problem in terms of the number of instructions $\mathcal{N}$ required by a Turing machine to solve the problem. If $\mathcal{N}$ is a polynomial in n the problem is termed "tractable," and if $\mathcal{N}$ grows faster than any polynomial in n the problem is termed "intractable". The addition of two n-digit numbers is a tractable problem, and the factorization of a number into primes is believed to be intractable, although there is no formal proof of this. Two important complexity classes are the **P** class, the class of problems which are tractable, and the **NP** class, that of problems whose solution, if one can be found, can be *checked* in a polynomial time. [13] **NP** stands for "nondeterministic polynomial," which means that the corresponding class of problems can be solved using a branching algorithm, with instructions "go to both 1 and 2." The number of branches grows exponentially, which is the reason why finding the solution requires a nonpolynomial time, while exploring a single branch to check a solution requires only polynomial time. Naturally, $\mathbf{P} \subset \mathbf{NP}$, and there exists the celebrated conjecture $\mathbf{P} \neq \mathbf{NP}$, which to this day remains unproven. Numerous complexity classes have been identified using the definition based on the computational model of Turing machines, but independent of the actual model provided that the model can be simulated in a polynomial time by a Turing machine. In particular, one has identified **NP complete** problems, such as the traveling salesman problem: finding a polynomial algorithm for one of the **NP complete** problems would automatically imply a polynomial solution for all **NP** problems.

Up to now we have only discussed calculable problems. The *Church–Turing thesis*, which is universally agreed upon but is by its nature impossible to prove, states that *the class of functions which can be calculated by a Turing machine corresponds exactly to the class of functions which one would naturally consider to be calculable using an algorithm*. There exist properly identified problems which are not calculable, for which it is known that no algorithm exists. An example is the halting problem of a Turing machine: the function which associates with any program run on a Turing machine (a finite series of symbols) a 0 or a 1 according to whether the machine stops or does not stop is not a calculable function. Quantum computers also seem to be covered by the Church–Turing

---

[13] For example, it is intractable to decompose a number into primes, but it is tractable to check the product if the primes are known.

thesis: the functions which quantum computers can calculate are *a priori* the same as those calculable by Turing machines.

We have stated that the simulation of any model of a classical algorithm can be done up to a polynomial time on a Turing machine, and this result has become a sort of basic "axiom" of the theory of algorithmic complexity. This is the strong version of the Church–Turing thesis, which is stated as follows: *any computational model can be simulated on a probabilistic Turing machine with at most a polynomial increase in the number of computational steps.* Quantum computers are important in that they make this strong version questionable. In fact, if factorization is an intractable problem (as suggested by experience but unproven), then the Shor algorithm contradicts this strong version. Using a quantum computer it is possible to decompose a number into primes by a number of steps which is a polynomial in n, whereas a classical computer can only do this in an exponential number of steps. The power of the quantum algorithms is due to the fact that they can explore at the same time all the branches of a nondeterministic algorithm. As we have seen in the case of the Shor algorithm, it is a constructive interference of the different branches which allows us to select the right result. Unfortunately, factorization is not an **NP complete** problem, so that its polynomial solution does not imply that of all **NP** problems.

5.10 Exercises

5.10.1 Justification of the circuits of Fig. 5.4

1. Justify the upper circuit of Fig. 5.4. Show that the action of the cNOT gate on the tensor product

$$\frac{1}{\sqrt{2}}(|0\rangle + |1\rangle) \otimes |0\rangle$$

gives an entangled state.

2. Let us assume that qubits are measured immediately *after* a cU gate. Show that the probabilities of finding the target qubit in the state $|0\rangle$ or $|1\rangle$ and its final states are the same as if the control bit were measured *before* the gate and the target bit were transformed or not according to whether the control bit was in the state $|0\rangle$ or $|1\rangle$. This observation allows the two-qubit gate cU to be replaced by a one-qubit gate acting on the target bit, which is an enormous technical simplification. However, it works only at the end of a calculation, not for an intermediate cU gate!

3. Show that the lower circuit of Fig. 5.4 constructs the Toffoli gate, with $U = \sqrt{\sigma_x}$.

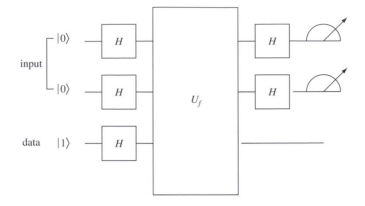

Figure 5.11 The Deutsch algorithm. The two qubits of the input register are initially in the state $|0\rangle$ and the qubit of the output register is in the state $|1\rangle$.

5.10.2 The Deutsch–Josza algorithm

The Deutsch algorithm (Section 5.4) can be generalized to the case where the input register contains two qubits and the output register contains only one qubit (Fig. 5.11). The two qubits of the input register are initially in the state $|0\rangle$, the qubit of the output register is in the state $|1\rangle$, and H is the Hadamard operator:

$$H|0\rangle = \frac{1}{\sqrt{2}} (|0\rangle + |1\rangle), \qquad H|1\rangle = \frac{1}{\sqrt{2}} (|0\rangle - |1\rangle).$$

The unknown function $f(x)$ is either

(i) $f(x) = $ constant, or
(ii) $f(x) = x \bmod 2$.

1. Show that the global state vector $|\Psi\rangle$ before entering the box U_f is

$$|\Psi\rangle = \left(\frac{1}{2} [|0 \otimes 0\rangle + |0 \otimes 1\rangle + |1 \otimes 0\rangle + |1 \otimes 1\rangle] \right) \otimes \left(\frac{1}{\sqrt{2}} [|0\rangle - |1\rangle] \right),$$

where the quantity in the first set of parentheses is the state vector of the two qubits of the input register and that in the second is the state vector of the qubit of the output register.

2. Let us recall the action of the box U_f (where x is the input register and y is the output register):

$$U_f|x \otimes y\rangle = |x \otimes [y \oplus f(x)]\rangle,$$

where $\oplus$ is mod 2 addition. Write down the state vector $U_f|\Psi\rangle$ in the cases (i) and (ii).

3. Write down the final state vector $|\Phi\rangle$ of the qubits of the input register (that is, *after* application of the operator $H \otimes H$) in cases (i) and (ii). Show that measurement of these qubits allows the cases (i) and (ii) to be distinguished.

5.10.3 Grover algorithm and constructive interference

Let us write the state vector after application of the Hadamard gates in Fig. 5.8

$$|\Psi\rangle = \frac{1}{\sqrt{N}} \sum_x a_x^{(0)} |x\rangle \quad a_x^{(0)} = 1.$$

Show that the application of the operator GO gives

$$GO|\Psi\rangle = \frac{1}{\sqrt{N}} \sum_x a_x^{(1)} |x\rangle$$

with

$$a_x^{(1)} = \frac{2}{N} \left(\sum_y (-1)^{f(y)} a_y^{(0)} \right) - (-1)^{f(x)} a_x^{(0)}.$$

Take, for example, $N = 16$ and show that $a_x^{(1)} = 3/4$ if $x \neq x_0$ and $a_x^{(1)} = 11/4$ if $x = x_0$. Constructive interference increases the probability for finding the final qubit state corresponding to $x = x_0$. Repeated application of $GO \sqrt{N}$ times allows the solution to stand out against a very small background.

5.10.4 Example of finding y_j

Let us take the example of Box 2.3, which shows that a possible period is $r = 3$. We choose $n = 4$, $2^n - 1 = 15$. What are the values of y_j such that $|\delta_j| < 1/2$ in (5.44)? Calculate the corresponding probability $p(y_j)$ and show that the sum of these probabilities is greater than 40%.

5.11 Further reading

A popularized approach to quantum computing can be found in Johnson (2003). A great deal of information about circuits and quantum algorithms can be found in Nielsen and Chuang (2000), Chapter 5, Stolze and Suter (2004), Chapters 3, 5 and 8, Preskill (1999), Chapter 6, and Mermin (2003). The Shor algorithm is discussed in detail in Ekert and Josza (1996). Interesting references on general problems in information theory are Bennett (1987) and Landauer (1991a) and (1991b).

6

Physical realizations

We are still at the very beginnings of physical implementations of quantum computers. The devices listed below have been used successfully to entangle two qubits (at most!), except for NMR which has gone up to 7 qubits. It is still premature to try to predict which device will prove most effective for building a quantum computer capable of dealing with several hundred qubits (if such a computer someday exists); perhaps it will be something new, not on this list at all. In any case, it would be as foolish to predict that such a computer will not be available by 2050 as to predict the contrary. [1]

The storage and processing of quantum information requires physical systems possessing the following properties (di Vincenzo criteria):

(i)   they must be scalable, that is, capable of being extended to a sufficient number of qubits, with well defined qubits;

(ii) they must have qubits which can be initialized in the state $|0\rangle$;

(iii) they must have qubits which are carried by physical states of sufficiently long lifetime, so as to ensure that the quantum states remain coherent throughout the calculation;

(iv) they must possess a set of universal quantum gates: unitary transformations on individual qubits and a cNOT gate, which are obtained by controlled manipulations;

(v)   there must be an efficient procedure for measuring the state of the qubits at the end of the calculation (*readout* of the results).

The Enemy Number One of a quantum computer is interaction with the environment leading, as we have seen in Section 4.4, to the phenomenon of decoherence, a consequence of which is the loss of the phase in the linear superposition of qubits. The calculations must be performed in a time less than the decoherence

---

[1] We note that the time scale here is similar to that foreseen for obtaining energy from fusion, a first step in this direction being the ITER project.

time τ_{D}. If an elementary operation (logic gate) on a qubit takes a time τ_{op}, the figure of merit for a quantum computer is the ratio

$$n_{\mathrm{op}} = \frac{\tau_{\mathrm{D}}}{\tau_{\mathrm{op}}}.$$

This is the maximum number of operations that the quantum computer can perform. The devices imagined up to now include the following (this list is not exhaustive):

- a photonic quantum computer based on the nonlinear Kerr effect;
- optical resonant cavities;
- microwave resonant cavities;
- ion traps;
- nuclear magnetic resonance;
- superconducting circuits with Josephson junctions;
- quantum dots;
- atoms of a Bose–Einstein condensate trapped in an optical lattice.

In this chapter we shall limit ourselves to four types of device: NMR (Section 6.1), trapped ions (Section 6.2), superconducting qubits (Section 6.3), and quantum dots (Section 6.4). The necessary background for understanding NMR quantum computers has been given in Section 3.4. However, Sections 6.2 to 6.4 require a more advanced knowledge of physics than has been assumed up to now in this book, even though we restrict ourselves to schematic descriptions. The reader can proceed directly to Chapter 7 (quantum information), which is completely independent of the present chapter.

6.1 NMR as a quantum computer

The record in the number of qubits was set in 2001 by a quantum computer using NMR. In spite of this record, NMR is certainly not a solution with a future, owing to problems which we shall discuss later on. As a preliminary, let us reformulate the results of Section 3.3 using a more abstract but also more general formalism, which will allow us in particular to treat easily the case of two coupled spins. The Hamiltonian (3.24) can be written as

$$\hat{H}(t) = \hat{H}_0 + \hat{H}_1(t) = -\frac{\hbar}{2}\omega_0 \sigma_z - \frac{\hbar}{2}\omega_1 \left(\sigma_+ e^{i\omega t} + \sigma_- e^{-i\omega t}\right) \qquad (6.1)$$

with $\sigma_\pm = (\sigma_x \pm i\sigma_y)/2$:

$$\sigma_+ = \begin{pmatrix} 0 & 1 \\ 0 & 0 \end{pmatrix}, \qquad \sigma_- = \sigma_+^\dagger = \begin{pmatrix} 0 & 0 \\ 1 & 0 \end{pmatrix}.$$

To study the evolution of the state vector $|\varphi(t)\rangle$ we define the state vector $|\tilde{\varphi}(t)\rangle$ in the "rotating reference frame" as

$$|\tilde{\varphi}(t)\rangle = \exp\left[-\frac{i\omega\sigma_z t}{2}\right] |\varphi(t)\rangle. \tag{6.2}$$

To interpret this reference frame physically, we note that $|\tilde{\varphi}(t)\rangle$ is independent of time for $\omega = \omega_0$ and $\omega_1 = 0$:

$$|\tilde{\varphi}(t)\rangle = |\tilde{\varphi}(t=0)\rangle \qquad \omega = \omega_0, \qquad \omega_1 = 0$$

because

$$|\varphi(t)\rangle = e^{-i\hat{H}_0 t/\hbar}|\varphi(0)\rangle = e^{i\omega\sigma_z t/2}|\varphi(0)\rangle \qquad \text{if} \qquad \omega = \omega_0. \tag{6.3}$$

This tells us that the spin remains at rest in the rotating reference frame, and so the expectation value $\langle\tilde{\varphi}(t)|\vec{\sigma}|\tilde{\varphi}(t)\rangle$ of $\vec{\sigma}$ is independent of time, while in the laboratory frame this expectation value rotates with angular velocity ω_0. In general, when $\omega \neq \omega_0$ the spin rotates in the frame (6.2) with angular velocity $(\omega_0 - \omega)$. We can easily obtain the evolution equation of $|\tilde{\varphi}(t)\rangle$ when $\omega_1 \neq 0$:

$$i\hbar\frac{d|\tilde{\varphi}\rangle}{dt} = \left(\frac{\hbar}{2}\delta\sigma_z - \frac{\hbar}{2}\omega_1\sigma_x\right)|\tilde{\varphi}(t)\rangle = \tilde{H}|\tilde{\varphi}(t)\rangle. \tag{6.4}$$

We recall that $\delta = \omega - \omega_0$ is the detuning defined in Fig. 3.4. The Hamiltonian $\tilde{H}$ has become independent of time in the rotating frame! In obtaining (6.4) we have used

$$\tilde{H}(t) = \frac{\hbar}{2}\omega\sigma_z + e^{-i\omega\sigma_z t/2}\,\hat{H}(t)\,e^{i\omega\sigma_z t/2}$$

and [2]

$$\tilde{\sigma}_\pm(t) = e^{-i\omega\sigma_z t/2}\,\sigma_\pm\,e^{i\omega\sigma_z t/2} = e^{\mp i\omega t}\sigma_\pm. \tag{6.5}$$

Now we can give a geometrical interpretation of the effect of the radiofrequency field in the rotating frame. To simplify matters let assume that $\omega = \omega_0$ and leave the general case to Exercise 6.5.1. Then

$$e^{-i\tilde{H}t/\hbar} = e^{i\omega_1\sigma_x t/2} = \cos\frac{\omega_1 t}{2} + i\sigma_x\sin\frac{\omega_1 t}{2}. \tag{6.6}$$

Since the operator [3] which rotates the spin by an angle θ about Ox is $R_x(\theta) = \exp(-i\theta\sigma_x/2)$, we see that $\exp(i\omega_1 t\sigma_x/2)$ is the operator which rotates by an

---

[2] The simplest way to find (6.5) is to note that $\tilde{\sigma}_\pm(t)$ satisfies the differential equation

$$\frac{d\tilde{\sigma}_\pm(t)}{dt} = -\frac{i\omega}{2}e^{-i\omega\sigma_z t/2}[\sigma_z, \sigma_\pm]e^{i\omega\sigma_z t/2} = \mp i\omega\tilde{\sigma}_\pm(t)$$

because $[\sigma_z, \sigma_\pm] = \pm 2\sigma_\pm$.

[3] We refer the reader to Exercise 3.5.1.

angle $\theta = -\omega_1 t$ about Ox. The spin can be made to rotate by a given angle by adjusting the duration t of the radiofrequency pulse. In particular, a $\pi/2$ pulse (Section 3.3) of duration $\omega_1 t/2 = \pi/4$ rotates the spin by $\pi/2$ about Ox: if the spin is initially along Oz, then this rotation aligns it with Oy.

The advantage of the preceding formalism is that it permits a convenient treatment of the case of two coupled spins, which we shall now study. In order to avoid a proliferation of indices, we use (X, Y, Z) to denote the Pauli matrices:

$$X \equiv \sigma_x, \qquad Y \equiv \sigma_y, \qquad Z \equiv \sigma_z. \tag{6.7}$$

Let us consider two spins $1/2$ attached to the *same* molecule, [4] for example, the first spin (1) carried by a proton and the second (2) carried by a ^{13}C nucleus. These two spins have different magnetic moments and therefore different resonance frequencies $\omega_0^{(1)}$ and $\omega_0^{(2)}$ and different Rabi frequencies $\omega_1^{(1)}$ and $\omega_1^{(2)}$. If the two spins are carried by identical nuclei, it is the chemical shift which causes the resonance frequencies to be different, but the difference will of course be very small in this case, $\sim 10^{-5}$ in relative value. The spins are coupled by an interaction [5] of the type $\hbar J Z_1 Z_2$ (more correctly, $\hbar J Z_1 \otimes Z_2$, but we shall frequently omit the tensor product notation: $Z_1 Z_2 \equiv Z_1 \otimes Z_2$, $Z_2 \equiv I_1 \otimes Z_2$, and so on). The Hamiltonian $\hat{H}_{12}(t)$ of the set of two spins is then obtained by generalizing (6.1):

$$\hat{H}_{12}(t) = -\frac{\hbar}{2} \omega_0^{(1)} Z_1 - \frac{\hbar}{2} \omega_0^{(2)} Z_2 - \frac{\hbar}{2} \omega_1^{(1)} \left(\sigma_{1+} e^{i\omega^{(1)}t} + \sigma_{1-} e^{-i\omega^{(1)}t} \right)$$
$$- \frac{\hbar}{2} \left(\omega_1^{(2)} \sigma_{2+} e^{i\omega^{(2)}t} + \sigma_{2-} e^{-i\omega^{(2)}t} \right) + \hbar J Z_1 Z_2 \tag{6.8}$$

with $\sigma_{i\pm} = (X_i \pm iY_i)/2$. Since the resonance frequencies are different, the fields applied to the two spins will have different frequencies, adjusted to be in quasi-resonance with each spin:

$$|\delta^{(1)}| = |\omega^{(1)} - \omega_0^{(1)}| \ll \omega_1^{(1)}, \qquad |\delta^{(2)}| = |\omega^{(2)} - \omega_0^{(2)}| \ll \omega_1^{(2)}. \tag{6.9}$$

In the rotating frame the state vector $|\tilde{\varphi}_1(t) \otimes \tilde{\varphi}_2(t)\rangle$ of the system of two spins is given by the generalization of (6.2):

$$|\tilde{\varphi}_1(t) \otimes \tilde{\varphi}_2(t)\rangle = \exp\left[-\frac{i\omega^{(1)} Z_1 t}{2} \right] \exp\left[-\frac{i\omega^{(2)} Z_2 t}{2} \right] |\varphi_1(t) \otimes \varphi_2(t)\rangle. \tag{6.10}$$

In this reference frame the Hamiltonian is as before independent of time:

$$\tilde{H} = \frac{\hbar}{2} \delta_1 Z_1 + \frac{\hbar}{2} \delta_2 Z_2 - \frac{\hbar}{2} \omega_1^{(1)} X_1 - \frac{\hbar}{2} \omega_1^{(2)} X_2 + \hbar J Z_1 Z_2, \tag{6.11}$$

---

[4] The molecules are diluted in a solvent, and the interactions between the "active" molecules, that is, those carrying the qubits, are negligible.

[5] This interaction is indirect and is not due to an interaction between the magnetic moments. It is transmitted by the electrons involved in the same chemical bond.

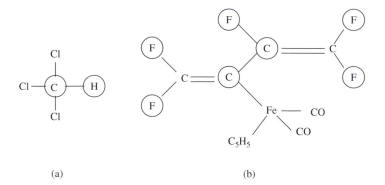

(a)   (b)

Figure 6.1 Two molecules used for quantum computing; the atoms carrying the active qubits are circled. (a) Chloroform, 2 qubits; (b) a perfluorobutadienyl iron complex, 7 qubits.

where we have used the fact that $Z_1 Z_2$ commutes with Z_1 and Z_2. In what follows it will be important to bear in mind that the condition $|J| \ll \omega_1^{(1)}, \omega_1^{(2)}$ is always satisfied in practice. Two examples of molecules which have been used successfully are given in Fig. 6.1.

Let us now consider the quantum logic gates. The qubits are of course spins 1/2; for the moment we shall ignore the fact that the qubits are in a complex environment, and proceed as though we were dealing with individual qubits. The manipulation of qubits one by one corresponding to one-qubit logic gates is obvious: it is sufficient to apply a radiofrequency field for a suitable time interval, where the frequency is close to the resonance frequency $\omega_0^{(i)}$ of the qubit (i) which we wish to manipulate.

The cNOT gate is realized using the interaction $JZ_1 \otimes Z_2$ between the two qubits. As we have seen, it is impossible to realize a cNOT gate by manipulating individual qubits. NMR differs from other quantum computer devices in that the interaction between qubits is not introduced externally, but is *internal to the system*. The interaction $JZ_1 \otimes Z_2$ between the spins is always present, and the problem is to suppress its effects when we want certain qubits not to evolve. We shall use the fact that the typical time needed for the $JZ_1 \otimes Z_2$ term to perform a rotation of two qubits (several milliseconds) is about two orders of magnitude larger than the time needed for the radiofrequency field to rotate an individual qubit (about ten microseconds). We can immediately calculate the evolution operator $\exp(-itJZ_1 \otimes Z_2)$ using

$$(Z_1 \otimes Z_2)(Z_1 \otimes Z_2) = Z_1^2 \otimes Z_2^2 = I_{12}.$$

We then find

$$\exp(-iJt\,Z_1 \otimes Z_2) = I_{12} \cos Jt - i(Z_1 \otimes Z_2) \sin Jt. \tag{6.12}$$

The following realization of a cNOT gate uses the operators performing $\pi/2$ rotations applied to individual qubits. The operators which rotate by an angle $\pi/2$ about Ox, Oy, and Oz, that is, $R_x(\pi/2)$, $R_y(\pi/2)$, and $R_z(\pi/2)$, are obtained using the fact that $\exp(-i\theta\vec{\sigma}\cdot\hat{n}/2)$ is the operator for a rotation $R_{\hat{n}}(\theta)$ by an angle θ about the $\hat{n}$ axis (Exercise 3.5.1). We then have

$$R_x(\pi/2) = \frac{1}{\sqrt{2}}(I - iX), \qquad R_y(\pi/2) = \frac{1}{\sqrt{2}}(I - iY), \qquad R_z(\pi/2) = \frac{1}{\sqrt{2}}(I - iZ).$$

$$(6.13)$$

We define the operator $X_{12}(t)$ acting on the ensemble of two spins

$$X_{12}(t) = \exp[iJt(Z_1 \otimes Z_2)]\, R_z^{(1)}(\pi/2)\, R_z^{(2)}(\pi/2)$$

during a time t such that $Jt = \pi/4$:

$$\exp[i\pi(Z_1 \otimes Z_2)/4] = \frac{1}{\sqrt{2}}(1 + iZ_1 \otimes Z_2).$$

Now we have

$$X_{12}\left(\frac{\pi}{4J}\right) = \left(\frac{1}{\sqrt{2}}\right)^3 (I_{12} + iZ_1 \otimes Z_2)(I_{12} - iZ_1 \otimes I_2)(I_{12} - iI_1 \otimes Z_2).$$

The multiplication can be done immediately, giving

$$X_{12}\left(\frac{\pi}{4J}\right) = \frac{1-i}{\sqrt{2}}(1 + Z_1 \otimes I_2 + I_1 \otimes Z_2 - Z_1 \otimes Z_2) = \frac{1-i}{\sqrt{2}}\, cZ, \qquad (6.14)$$

where the cZ (control-Z) gate is

$$cZ = \begin{pmatrix} 1 & 0 & 0 & 0 \\ 0 & 1 & 0 & 0 \\ 0 & 0 & 1 & 0 \\ 0 & 0 & 0 & -1 \end{pmatrix} = \begin{pmatrix} I & 0 \\ 0 & \sigma_z \end{pmatrix}.$$

To go from the cZ gate to the cNOT = cX gate, it is sufficient to sandwich the former between two Hadamard gates acting on qubit 2:

$$cNOT = (I_1 \otimes H_2)\, cZ\, (I_1 \otimes H_2). \qquad (6.15)$$

In fact,

$$\begin{pmatrix} H & 0 \\ 0 & H \end{pmatrix} \begin{pmatrix} I & 0 \\ 0 & \sigma_z \end{pmatrix} \begin{pmatrix} H & 0 \\ 0 & H \end{pmatrix} = \begin{pmatrix} H^2 & 0 \\ 0 & H\sigma_z H \end{pmatrix} = \begin{pmatrix} I & 0 \\ 0 & \sigma_x \end{pmatrix}.$$

The Hadamard gate corresponds to a rotation of π about the axis $(1/\sqrt{2}, 0, 1/\sqrt{2})$:

$$\exp\left(\frac{-i\pi(\sigma_x + \sigma_z)}{2\sqrt{2}}\right) = -iH,$$

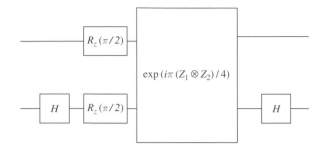

Figure 6.2 NMR construction of a cNOT gate. The diagrams are read from left to right, and the operator products are taken from right to left.

but in practice rotations about Ox or Oy are always used. The time needed for the rotations $R_z(\pi/2)$ or H (about ten microseconds) is negligible compared to the time needed for the evolution due to the term $JZ_1 \otimes Z_2$ (several milliseconds), and this evolution is negligible during the time needed for the individual rotations. The logic circuit corresponding to the operations (6.14) and (6.15) is shown in Fig. 6.2.

However, during the several milliseconds needed to realize the gate, the other qubits continue to evolve according to the various terms in the Hamiltonian. The NMR signal is not produced by a single spin, but by an ensemble of spins ($\sim 10^{18}$ of them, the minimum number needed to obtain a measurable signal). The nonuniformities of the field $\vec{B}_0$ and other random phenomena cause the qubits carried by different molecules to evolve differently, and so the signal will become fuzzy. This is why it is necessary to resort to the technique of *refocusing* (spin echo), a basic tool used in modern NMR and MRI. We shall explain it here for the simplified case of evolution due only to the $JZ_1 \otimes Z_2$ term. Sandwiching the evolution operator between two rotations $R_x^{(1)}(\pi) = -iX_1$ of spin 1, we obtain the following effect:

$$(-iX_1)[\exp(-iJt\, Z_1 \otimes Z_2)](-iX_1) = (-iX_1)(I_{12} \cos Jt - i(Z_1 \otimes Z_2)\sin Jt)(-iX_1)$$

$$= I_{12} \cos Jt + i(Z_1 \otimes Z_2)\sin Jt \qquad (6.16)$$

$$= \exp(+iJt\, Z_1 \otimes Z_2).$$

Therefore, if the spins have evolved during a time t as $\exp(-iJt\, Z_1 \otimes Z_2)$, we obtain the result

$$R_x^{(1)}(\pi)\,[\exp(-iJtZ_1 \otimes Z_2)]\,R_x^{(1)}(\pi)\,[\exp(-iJtZ_1 \otimes Z_2)] = I_{12}. \qquad (6.17)$$

In other words, the sequence of operations corresponding to free evolution during a time $t \times R_x^{(1)}(\pi) \times$ time evolution during a time $t \times R_x^{(1)}(\pi)$ puts the spins back in their initial configuration! This observation shows how it is possible to cancel out the evolution of the qubits other than those to which the cNOT gate is applied by means of the operations (6.14) and (6.15). A similar observation reveals how spins which have evolved differently owing to nonuniformities of $\vec{B}_0$ can be refocused.

The molecule (b) of Fig. 6.1 allows us to work with 7 qubits, the minimum number of qubits needed to use the Shor algorithm to factorize 15 into primes. In fact, b can take the values 2, 4, 7, 8, 11, 13, or 14, and the largest period of $b^x \bmod N$ is $r = 4$, for $b = 2$, $b = 7$, $b = 8$ and $b = 13$. To see two periods it is then necessary to take $x = 0, 1, \ldots, 7 = 2^3 - 1$, and of course $f(x) = 0, 1, \ldots, 15 = 2^4 - 1$, that is, a 3-qubit input register and a 4-qubit output register. The factorization of 15 was successfully done in 2001, thanks to the great sophistication of the modern NMR techniques developed for chemical and biological analyses. However, despite this spectacular (?) result, NMR is not the solution of the future, because it first requires the synthesis of a molecule possessing as many distinguishable sites as the needed number of qubits, and the ability to select the frequencies acting on the various qubits. Worst of all, the signal decreases exponentially as the number of qubits grows. In fact, NMR uses not individual quantum objects, but a set of $\gtrsim 10^{18}$ active molecules diluted in a solvent: the signal is a *collective* one. To obtain a "pseudo-pure" state it is necessary to perform preliminary initialization operations too complex to be described here, and the same for measuring the final states. It is these operations which cause the signal to fall off as the number of qubits is increased.

6.2 Trapped ions

Trapped ions represent a more promising technique than NMR. The two states of a qubit are carried by the ground state $(|g\rangle \equiv |0\rangle)$ of an ion and by an excited state of very long (~ 1 s) lifetime $|e\rangle \equiv |1\rangle$, which is either a hyperfine state of the electronic ground state, or a metastable electronic state. These states will be called *internal states* of the ions. The individual qubits are manipulated by laser pulses, as explained in Section 3.3. The construction of entangled states and two-qubit logic gates involves as an intermediary the translational motion, called the *external degrees of freedom* of the ions. A coupling between the internal and external degrees of freedom is therefore used. Since the ions are trapped in a harmonic potential, we shall speak of vibrational motion of the ions rather than translational motion.

The traps which are used, named Paul traps after their inventor, are constructed by combining the actions of constant and alternating electric fields. The net result is that the ions are located in a harmonic potential

$$V(x, y, z) = \frac{1}{2} M \left(\omega_x^2 x^2 + \omega_y^2 y^2 + \omega_z^2 z^2 \right),$$

where M is the ion mass and $\vec{r} = (x, y, z)$ is the position of the ion in the trap. In practice, the trap frequencies, typically a few megahertz, satisfy the equation

$$\omega_x^2 \sim \omega_y^2 \gg \omega_z^2,$$

so that in a first approximation the ion moves along the axis Oz in a potential

$$V(z) = \frac{1}{2} M \omega_z^2 z^2. \tag{6.18}$$

To start with an elementary discussion, it is useful first to study the case of a single trapped ion. In quantum physics, the z coordinate and the z component of the momentum, p_z, are Hermitian operators (physical properties) satisfying the commutation relation

$$[z, p_z] = i\hbar I. \tag{6.19}$$

The Hamiltonian $\hat{H}$ with the potential energy term (6.18) also contains a kinetic energy term $p_z^2 / 2M$:

$$\hat{H} = \frac{p_z^2}{2M} + \frac{1}{2} M \omega_z^2 z^2. \tag{6.20}$$

There exists a standard method for finding the eigenvalues (energy levels) and eigenvectors (stationary states) of $\hat{H}$. One introduces the (dimensionless) operator a and its Hermitian conjugate $a^\dagger$:

$$a = \sqrt{\frac{M\omega_z}{2\hbar}} \left(z + \frac{i p_z}{M\omega_z} \right), \qquad a^\dagger = \sqrt{\frac{M\omega_z}{2\hbar}} \left(z - \frac{i p_z}{M\omega_z} \right). \tag{6.21}$$

Using (6.19), it can immediately be checked (Exercise 6.5.2) that a and $a^\dagger$ satisfy the commutation relation

$$\left[a, a^\dagger \right] = I, \tag{6.22}$$

and that $\hat{H}$ can be rewritten as

$$\hat{H} = \hbar \omega_z \left(a^\dagger a + \frac{1}{2} \right). \tag{6.23}$$

It is shown in any quantum mechanics text that the eigenvalues of $\hat{H}$ are of the form $\hbar \omega_z (m + 1/2)$, $m = 0, 1, 2, \ldots$, corresponding to the eigenvectors $|m\rangle$:

$$\hat{H}|m\rangle = \hbar \omega_z \left(m + \frac{1}{2} \right) |m\rangle. \tag{6.24}$$

In general, the operator a (the annihilation operator) takes m to $m-1$, and the operator $a^\dagger$ (the creation operator) takes m to $m+1$:

$$a|m\rangle = \sqrt{m}\,|m-1\rangle, \qquad a^\dagger|m\rangle = \sqrt{m+1}\,|m+1\rangle. \tag{6.25}$$

The integer m therefore labels the vibrational quantum number in the trap. According to (6.25), the ground state $|0\rangle$ is "annihilated" by a: $a|0\rangle = 0$. Its energy E_0 is nonzero: $E_0 = \hbar\omega_z/2$, in contrast to the classical case where the ground state corresponds to an ion at rest in the bottom of the potential well at $z = 0$. The fact that the ground-state energy is nonzero has an interesting physical interpretation in terms of the Heisenberg inequalities (Exercise 2.6.5). Reasoning heuristically, we can replace z and p_z by their dispersions Δz and Δp_z, and use the Heisenberg inequality in the form $\Delta z\,\Delta p_z \sim \hbar/2$ in (6.20) to obtain

$$E \sim \frac{(\Delta p_z)^2}{2M} + \frac{1}{2}M\omega^2(\Delta z)^2 \sim \frac{1}{8M(\Delta z)^2} + \frac{1}{2}M\omega^2(\Delta z)^2.$$

Minimizing with respect to Δz, we find

$$(\Delta z)^2 = \frac{\hbar}{2M\omega_z}, \qquad E_0 = \frac{\hbar}{2}\omega_z, \tag{6.26}$$

in agreement (accidentally – we expected to obtain only the correct order of magnitude) with the exact calculation. This heuristic argument shows that the ground-state energy is obtained by seeking the best compromise between the kinetic energy and the potential energy. They cannot both vanish as in the classical case. The argument also shows that the spread of the ion wave function in the trap, that is, the region where the ion has an appreciable probability of being located, is of order $\Delta z_0 = \sqrt{\hbar/2M\omega_z}$. It is usual to redefine the zero of the vibrational energy such that the ground state has zero energy. Then the energies take the simple form $m\hbar\omega_z$.

There is one final experimental condition to be satisfied. In order to be able to manipulate the ion, it must be in its vibrational ground state $m = 0$. This will not be the case if the ion is at a temperature T such that $k_B T \gtrsim \hbar\omega_z$. In that case, levels with $m \neq 0$ will be thermally excited, and it becomes essential to cool the ions. This is done by Doppler cooling based on the following principle. The ion is sandwiched between two laser beams propagating in opposite directions[6] and tuned to slightly below a resonance frequency. When an ion travels opposite to the direction of one of the laser beams, the transition becomes closer to resonance owing to the Doppler effect, because the ion "sees" more energetic photons, and the absorption of photons from this beam becomes more important than that of photons from the second beam, which the ion "sees" as farther from resonance.

---

[6] For simplicity, we take the one-dimensional case. Cooling in three dimensions would require six laser beams.

The ion is therefore slowed down no matter what the direction of its velocity is, and it can be shown that the temperature the ion reaches is given by $k_B T \simeq \hbar\Gamma$, where Γ is the linewidth. If the Doppler cooling is insufficient, other more sophisticated mechanisms can be used.

First let us model the ion by a two-level system trapped in the potential (6.18) and located in an oscillating electric field:

$$\vec{E} = E_1 \hat{x} \cos(\omega t - kz - \phi). \tag{6.27}$$

Under these conditions, the total Hamiltonian contains three terms. The first, $\hat{H}_0$, is the Hamiltonian in the absence of the oscillating field ($E_1 = 0$):

$$\hat{H}_0 = -\frac{\hbar}{2}\omega_0 \sigma_z + \hbar\omega_z a^\dagger a. \tag{6.28}$$

The internal states are the two states $|0\rangle$ of energy $-\hbar\omega_0/2$ and $|1\rangle$ of energy $\hbar\omega_0/2$. We shall use the Hamiltonian (6.28) to define a "rotating frame," generalizing what we did in the NMR case.[7] Given an operator A, the operator $\tilde{A}(t)$ will be, by definition,

$$\tilde{A}(t) = e^{i\hat{H}_0 t/\hbar}\, A\, e^{-i\hat{H}_0 t/\hbar}. \tag{6.29}$$

Following the method of the preceding section,[8] we easily find for the operators a, $a^\dagger$, σ_-, and σ_+ (Exercise 6.5.2):

$$\tilde{\sigma}_+(t) = \sigma_+ e^{-i\omega_0 t}, \qquad \tilde{\sigma}_-(t) = \sigma_- e^{i\omega_0 t},$$
$$\tilde{a}(t) = a\, e^{-i\omega_z t}, \qquad \tilde{a}^\dagger(t) = a^\dagger e^{i\omega_z t}. \tag{6.30}$$

According to (6.27), the interaction with the electric field is written as

$$\hat{H}_{\text{int}} = -\frac{\hbar}{2}\omega_1 \left[\sigma_+ + \sigma_-\right]\left[e^{i(\omega t - \phi)}\, e^{-ikz} + e^{-i(\omega t - \phi)}\, e^{ikz}\right],$$

where ω_1 is the Rabi frequency of the problem. In this equation z is the position *operator*. We expand $\exp(\pm ikz)$ in a series keeping only the first two terms:

$$e^{\pm ikz} \simeq 1 \pm ikz,$$

which is valid if $k\Delta z_0 \ll 1$, where $\Delta z_0 = \sqrt{\hbar/2M\omega_z}$ is the spread of the ground-state wave function $m = 0$ in the trap. The condition for the expansion to be valid then is

$$\frac{k\sqrt{\hbar}}{\sqrt{2M\omega_z}} = k\Delta z_0 = \eta \ll 1,$$

---

[7] The reader familiar with quantum mechanics will recognize (6.29) as the definition of the interaction picture.
[8] However, we choose ω_0 as the rotational frequency of the rotating frame. In fact, it is more convenient to use $\hat{H}_0$ in this problem.

where η is called the *Lamb–Dicke* parameter. The term 1 in the expansion of $\exp(\pm ikz)$ gives a contribution $\hat{H}_1$ to $\hat{H}_{int}$:

$$\hat{H}_1 = -\frac{\hbar}{2}\omega_1 \left[\sigma_+ + \sigma_-\right]\left[e^{i(\omega t - \phi)} + e^{-i(\omega t - \phi)}\right].$$

In the rotating frame using the first line of (6.30) we find

$$\hat{H}_1 \to \tilde{H}_1 = -\frac{\hbar}{2}\omega_1 \left[\sigma_+ e^{-i\omega_0 t} + \sigma_- e^{i\omega_0 t}\right]\left[e^{i(\omega t - \phi)} + e^{-i(\omega t - \phi)}\right].$$

Finally, we use the *rotating-wave approximation*, where we neglect the terms in $\exp[\pm i(\omega + \omega_0)t]$ which oscillate rapidly so that they average to zero and give a negligible contribution to the evolution. This leads to the final form:

$$\tilde{H}_1 \simeq -\frac{\hbar}{2}\omega_1 \left[\sigma_+ e^{i(\delta t - \phi)} + \sigma_- e^{-i(\delta t - \phi)}\right] \tag{6.31}$$

with $\delta = (\omega - \omega_0)$. This is the NMR Hamiltonian (6.4) in the rotating frame, where we have included additional phase factors $\exp(\pm i\phi)$. It allows us to manipulate the two ion states exactly as in the case of NMR, by tuning the frequency of the oscillating field to $\omega = \omega_0$ ($\delta = 0$) and adjusting the duration of the interaction. The angle defining the rotational axis in the xOy plane can be chosen using the phase ϕ. In fact, at resonance ($\delta = 0$) the rotation operator is, according to (6.31) with $\theta = -\omega_1 t$,

$$\exp\left(-i\frac{\theta}{2}\left[\sigma_+ e^{-i\phi} + \sigma_- e^{i\phi}\right]\right) = \exp\left(-i\frac{\theta}{2}\left[\sigma_x \cos\phi + \sigma_y \sin\phi\right]\right),$$

which gives a rotation about the axis $\hat{n}$ with the components

$$\hat{n}_x = \cos\phi, \qquad \hat{n}_y = \sin\phi, \qquad \hat{n}_z = 0.$$

In fact, the value of the angle ϕ obviously has no absolute meaning, but in a series of several successive pulses it is the relative phases of the various pulses which are important. In what follows, we can choose an arbitrary value for ϕ. We have kept ϕ explicitly to make the link with Exercise 6.5.3, where it plays an essential role.

The term $\pm ikz$ in the expansion of the exponential $\exp(\pm ikz)$ gives a contribution $\hat{H}_2$ to the interaction Hamiltonian. This term takes into account the vibrational motion and couples the internal and external degrees of freedom:

$$\hat{H}_2 = \frac{i\hbar\eta\omega_1}{2}\left[\sigma_+ + \sigma_-\right]\left[a + a^\dagger\right]\left[e^{i(\omega t - \phi)} - e^{-i(\omega t - \phi)}\right], \tag{6.32}$$

where we have used (6.21) in the form

$$z = \sqrt{\frac{\hbar}{2M\omega_z}}\left(a + a^\dagger\right).$$

In the rotating frame the Hamiltonian $\hat{H}_2$ becomes

$$\hat{H}_2 \to \tilde{H}_2 = \frac{\mathrm{i}\eta\hbar\omega_1}{2}\left[\sigma_+ a\,\mathrm{e}^{-\mathrm{i}(\omega_0+\omega_z)t} + \sigma_+ a^\dagger\,\mathrm{e}^{-\mathrm{i}(\omega_0-\omega_z)t}\right.$$

$$\left. + \sigma_- a\,\mathrm{e}^{\mathrm{i}(\omega_0-\omega_z)t} + \sigma_- a^\dagger\,\mathrm{e}^{\mathrm{i}(\omega_0+\omega_z)t}\right]$$

$$\times \left[\mathrm{e}^{\mathrm{i}(\omega t-\phi)} - \mathrm{e}^{-\mathrm{i}(\omega t-\phi)}\right].$$

If we choose to tune the laser frequency to $\omega = (\omega + \omega_z)$, the so-called blue side-band frequency, $\tilde{H}_2$ becomes the following in the rotating-wave approximation:

$$\hat{H}_2 \to \tilde{H}_2 = \frac{\mathrm{i}\eta\,\hbar\omega_1}{2}\left[\sigma_+ a\,\mathrm{e}^{-\mathrm{i}\phi} - \sigma_- a^\dagger\mathrm{e}^{\mathrm{i}\phi}\right], \tag{6.33}$$

whereas if we choose $\omega = (\omega_0 - \omega_z)$, the so-called red side-band frequency, we have

$$\hat{H}_2 \to \tilde{H}_2 = \frac{\mathrm{i}\eta\,\hbar\omega_1}{2}\left[\sigma_+ a^\dagger\mathrm{e}^{-\mathrm{i}\phi} - \sigma_- a\,\mathrm{e}^{\mathrm{i}\phi}\right]. \tag{6.34}$$

We use $|n, m\rangle$ to denote the ion state, where $n = 0, 1$ is the internal state and $m = 0, 1$ is the vibrational state of the harmonic oscillator. The Hamiltonian (6.33) induces transitions between the states $|0, 0\rangle$ and $|1, 1\rangle$, because

$$\omega = \omega_0 + \omega_z: \qquad \sigma_+ a|1, 1\rangle = |0, 0\rangle, \qquad \sigma_- a^\dagger|0, 0\rangle = |1, 1\rangle,$$

whereas (6.34) induces transitions between the states $|0, 0\rangle$ and $|1, 1\rangle$, because

$$\omega = \omega_0 - \omega_z: \qquad \sigma_+ a^\dagger|1, 0\rangle = |0, 1\rangle, \qquad \sigma_- a|0, 1\rangle = |1, 0\rangle.$$

This is summarized by the level scheme shown in Fig. 6.3(a).

In order to explain in a simple way how (6.33) and (6.34) can lead to the formation of entangled states, it is convenient to assume the existence of an auxiliary *internal* state $|2\rangle$. It is possible to do without this auxiliary state, but then the discussion is a bit more complicated; see Exercise 6.5.3. We use $|n, m\rangle$ to denote the ion state, $n = 0, 1, 2$ being the internal (spin) state and m the excitation state of the harmonic oscillator. We then obtain the level scheme of Fig. 6.3(b). The four basis states for the quantum calculation are $|0, 0\rangle$, $|0, 1\rangle$, $|1, 0\rangle$ and $|1, 1\rangle$, and we need to construct two-qubit logic gates for these states. A laser is tuned to the frequency $(\omega_{\mathrm{aux}} + \omega_z)$, thus stimulating transitions between the states $|2, 0\rangle$ and $|1, 1\rangle$, corresponding to an effective Hamiltonian

$$\hat{H}_{\mathrm{aux}} = \mathrm{i}\frac{\eta\,\hbar\omega_1'}{2}\left[\sigma_+' a\,\mathrm{e}^{\mathrm{i}\phi} - \sigma_-' a^\dagger\mathrm{e}^{-\mathrm{i}\phi}\right] \tag{6.35}$$

with

$$\sigma_+'|1\rangle = |2\rangle, \qquad \sigma_-'|2\rangle = |1\rangle.$$

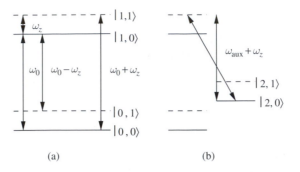

(a)   (b)

Figure 6.3 (a) Energy levels of the simplified model for the coupling of the
internal and external degrees of freedom. The states are labeled $|n, m\rangle$, where
$n = 0, 1$ is the state of the qubit (internal) and m is the vibrational quantum
number (external). (b) Coupling to an auxiliary level: now n can take the values
0, 1, and 2, with m always denoting the vibrational quantum number. The
transition at $\omega_0 - \omega_z$ corresponds to the red side-band, and that at $\omega_0 + \omega_z$ to the
blue side-band.

The laser is applied during the time needed to perform a rotation $R_x(2\pi)$, the
effect of which is $|1, 1\rangle \rightarrow -|1, 1\rangle$ with the other states remaining unchanged.
This realizes the cZ logic gate on the states $|n, m\rangle$:

$$cZ = \begin{pmatrix} 1 & 0 & 0 & 0 \\ 0 & 1 & 0 & 0 \\ 0 & 0 & 1 & 0 \\ 0 & 0 & 0 & -1 \end{pmatrix} = \begin{pmatrix} I & 0 \\ 0 & \sigma_z \end{pmatrix}. \tag{6.36}$$

Let us now turn to the more interesting case of two ions. The arguments can be
generalized immediately to any number of ions N, which allows us (in theory!) to
imagine a quantum computer with N qubits. As a preliminary result, we need the
SWAP gate, obtained by tuning the laser to the red side-band frequency ($\omega - \omega_z$)
and by adjusting the duration of the pulse for a rotation by π. Choosing $\phi = \pi/2$,
this gives the exchange $|0, 1\rangle \leftrightarrow |1, 0\rangle$ corresponding to the SWAP logic gate,
with an extra minus sign

$$SWAP' = \begin{pmatrix} 1 & 0 & 0 & 0 \\ 0 & 0 & -1 & 0 \\ 0 & 1 & 0 & 0 \\ 0 & 0 & 0 & 1 \end{pmatrix}. \tag{6.37}$$

The vibrational ground state corresponds to motion of the ensemble of both ions,
that is, to vibration of the center of mass inside the trap. Therefore, everything
is the same as for a single ion. The combination of cZ, SWAP, and Hadamard
gates gives a cNOT gate. Let us explain how. We choose ion 1 as the control ion
and ion 2 as the target ion. It should be borne in mind that the qubits are carried

by the two internal states of these ions. We start from a state which is the tensor product of the most general two-qubit state

$$a|00\rangle + b|01\rangle + c|10\rangle + d|11\rangle$$

and of the state corresponding to the $m = 0$ vibrational mode of the center of mass:

$$\text{initial} : (a|00\rangle + b|01\rangle + c|10\rangle + d|11\rangle) \otimes |0\rangle$$

$$= a|00, 0\rangle + b|01, 0\rangle + c|10, 0\rangle + d|11, 0\rangle$$

$$\text{SWAP}'_2 : a|00, 0\rangle + b|00, 1\rangle + c|10, 0\rangle + d|10, 1\rangle$$

$$\text{cZ}_1 : a|00, 0\rangle + b|00, 1\rangle + c|10, 0\rangle - d|10, 1\rangle$$

$$\text{SWAP}'_2 : a|00, 0\rangle - b|01, 0\rangle + c|10, 0\rangle + d|11, 0\rangle$$

$$= (a|00\rangle - b|01\rangle + c|10\rangle + d|11\rangle) \otimes |0\rangle.$$

If we redefine the phase of the state $|1\rangle_2$ of the second ion, $|1\rangle_2 \rightarrow -|1\rangle_2$, the net result is the application of a cZ gate to the two qubits: the vibrational motion has only served as an intermediary. It is easy to go from a cZ gate to a cNOT gate, as we have seen in (6.15). This gate has been realized experimentally using as qubits the ground state $S_{1/2}$ ($|g\rangle \equiv |0\rangle$) and the metastable state $D_{5/2}$ ($|e\rangle \equiv |1\rangle$, of lifetime of the order of a second) of the $^{40}\text{Ca}^+$ ion; the transition between the two levels is an electric quadrupole transition corresponding to a wavelength of 729 nm. An N-qubit computer is shown schematically in Fig. 6.4.

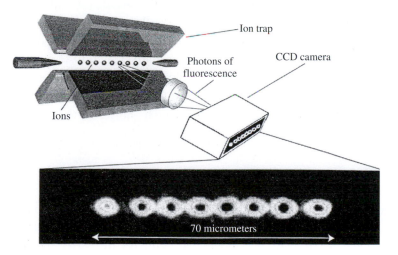

Figure 6.4 Schematic depiction of the principle of a quantum computer using N trapped ions. From Aspect and Grangier (2004).

The potential energy V of the ensemble of N ions in the trap is

$$V = \frac{1}{2} M \sum_{n=1}^{N} \omega_z^2 z_n^2 + \frac{q^2}{4\pi\varepsilon_0} \sum_{m \neq n} \frac{1}{|z_n - z_m|}, \tag{6.38}$$

assuming that the ion chain is linear: the trap potential must be sufficiently confining in the directions Ox and Oy in order to avoid zig-zag configurations. The minimum distance between the ions at equilibrium, which is the distance between the two central ions, is approximately

$$\Delta z \simeq 2 l N^{-0.057},$$

where l is the characteristic length of the problem:

$$l = \left(\frac{q^2}{4\pi\varepsilon_0 M \omega_z^2} \right)^{1/3}. \tag{6.39}$$

For the trap of the Innsbruck group the numerical value is $l \simeq 2.8\,\mu m$, the central ions being separated by about $5\,\mu m$. The lowest vibrational mode of frequency ω_z corresponds to motion of the ensemble of ions, and the first excited mode, or the breathing mode of frequency $\sqrt{3}\omega_z$, corresponds to the ions oscillating with amplitude proportional to their algebraic distance from the center of the trap (Exercise 6.5.4). One of the delicate problems is how to address an individual ion by a laser beam – it is necessary to aim very accurately!

A resonance fluorescence technique is used to put the ions in the desired state and to measure their final state $|g\rangle := |0\rangle$ or $|e\rangle := |1\rangle$. The ions are illuminated by a laser beam tuned to an electric dipole transition between the level $|g\rangle$ and an excited level $|r\rangle$, $|g\rangle \leftrightarrow |r\rangle$. If the ion is in the state $|g\rangle$, it will scatter a large number of photons, while if it is in the state $|e\rangle$ it will not. The quantum jumps made by the ions in this method are quite spectacular, because the observation is made for an *individual* quantum system. One main difficulty of trapped ions is that ions are charged particles, and as such are sensitive to stray electric fields. If these fields are time dependent, they will heat the ions. Typical heating times are of the order of 1 ms, but this time could be much shorter as the number of ions increases.

6.3 Superconducting qubits

In the two preceding examples the qubits were carried by individual quantum objects, nuclear spins in the NMR case and ions in the case of trapped ions (although in the NMR case the signal was built up by $\sim 10^{18}$ nuclear spins). We now turn to a system where the qubits are carried by a macroscopic degree of freedom, the current in a superconducting circuit containing one or several

Josephson junctions. As we shall see, low temperatures of the order of tens of millikelvins are required for these circuits to exhibit quantum behavior. These circuits are small by everyday standards (a few micrometers), but very large compared to atomic sizes. Still more remarkable is the fact that the parameters of these quantum systems are fixed by fabrication, and not by Nature as is the case for individual quantum systems like electrons or ions. They are engineered quantities which can be modified by changing the dimensions of the circuits, and in this sense they are unambiguously macroscopic quantities. It has been known for almost a century that at low temperatures the electrical resistance of most metals and alloys drops abruptly to zero below a transition temperature T_C of order 1 K, and the metal becomes a superconductor. Superconductors also exhibit a remarkable feature called the Meissner effect: magnetic fields are expelled from the bulk of a superconductor; they cannot penetrate deeper than a distance known as the London penetration length,[9] which is of order $0.1\,\mu$m.

In order first to examine a simple example, which, however, will turn out not to be suitable for qubits, consider the LC circuit of Fig. 6.5 at very low temperatures, so that all metallic elements are superconducting. The classical Hamiltonian of the oscillator is the sum of the magnetic energy stored in the inductor and the electrical energy stored in the capacitor:

$$H = \frac{1}{2L}\varphi^2 + \frac{1}{2C}q^2, \tag{6.40}$$

where L and C are the inductance and the capacitance, φ is the flux across the inductor, and q is the charge on the capacitor. The resonance frequency of the circuit is $\omega_0 = \sqrt{LC}$. The circuit of Fig. 6.5 can be fabricated with lateral

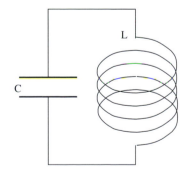

Figure 6.5 A superconducting LC circuit.

---

[9] We limit ourselves to the so-called type I superconductors, or to type II superconductors in external magnetic fields smaller than the critical field H_{C_1}. We also exclude from our discussion high-T_C superconductors, where the transition temperature can reach $\sim$100 K. In contrast to low-temperature superconductors, which are very well described by the Bardeen–Cooper–Schrieffer (BCS) theory, high-T_C superconductors are still poorly understood.

dimensions $\simeq 10\,\mu$m, with values of L and C approximately $0.1\,$nH and $1\,$pF, respectively, corresponding to a resonance frequency $\omega_0/2\pi \simeq 16\,$GHz. When aluminum with $T_C = 1.1\,$K is used to fabricate the circuit one can safely neglect dissipation due to unpaired electrons (quasi-particles) below [10] $20\,$mK.

The magnetic flux φ and the charge q in (6.40) may be considered as conjugate variables in the sense of analytical mechanics: the Hamiltonian (6.40) has the same form as (6.20) if we make the substitutions $p_z \to \varphi$, $M \to L$, $z \to q$, and $M\omega_z^2 \to 1/C$. The correspondence principle tells us how to quantize the circuit. As in the case of the variables z and p_z, the classical variables (numbers) q and φ become operators Q and Φ which obey the canonical commutation relation $[Q, \Phi] = i\hbar I$. However, we know that under ordinary conditions quantum effects are (fortunately) quite negligible in electrical circuits. So when do quantum effects begin to play a role? As in the case of trapped ions, the thermal energy $k_B T$ must be much smaller than the energy difference between the ground and first excited states. According to the results described in the preceding section on the quantum harmonic oscillator, here this energy difference is $\hbar\omega_0$, and quantum effects are relevant for the circuit of Fig. 6.5 when

$$k_B T \ll \hbar\omega_0. \tag{6.41}$$

Unfortunately, this simple circuit is not suitable as a support for qubits. The reason is that the energy difference between the first and second excited states is also $\hbar\omega_0$, so that any attempt to produce Rabi oscillations between the ground and first excited levels will unavoidably induce transitions to the second excited and higher levels, so that there is no possibility of a two-level system. The equal spacing $\hbar\omega_0$ between the harmonic oscillator quantum levels can be traced back to the linearity of the oscillator. We thus need to introduce strong nonlinear effects, and the only known device which is able to create strong nonlinearities without dissipation is the Josephson junction.

In a superconductor the electrons are weakly bound in pairs, called Cooper pairs, of zero spin and zero momentum at equilibrium. At zero temperature all these pairs "condense," that is, they all fall into the same state, the ground state, so that all pairs have the same wave function. [11] The energy needed to break one of the pairs is called the gap energy Δ, and it is the energy difference between the ground and first excited states. In contrast to a single-electron wave function, which is a probability amplitude whose modulus squared gives the *probability*

---

[10] As explained below, the characteristic energy is the gap energy Δ, and dissipation due to unpaired electrons is negligible if $k_B T \ll \Delta$, because the residual resistance decreases as $\sim \exp(-\Delta/k_B T)$. The value of the gap for aluminum is $\Delta \simeq 200\,\mu$eV, and the number of quasi-particles in a typical circuit at $20\,$mK is less than 10^{-10}.

[11] This phenomenon is known as the Bose–Einstein condensation of bosons. Electrons are fermions, not bosons, but Cooper *pairs* do behave as bosons.

density of finding the electron at some point in space, the wave function of the condensed state, as it involves a macroscopic number of pairs, may be interpreted in terms of a Cooper pair density $\rho(\vec{r})$. More precisely, this macroscopic wave function is a complex function which can be written as

$$\psi(\vec{r}) = \sqrt{\rho(\vec{r})}\, e^{i\theta(\vec{r})}, \tag{6.42}$$

so that $|\psi(\vec{r})|^2 = \rho(\vec{r})$ is the Cooper pair density at the point $\vec{r}$ and $\theta(\vec{r})$ is the phase. In a uniform situation ρ and θ are independent of $\vec{r}$. The Meissner effect is explained by starting from the standard form of the electromagnetic current in quantum mechanics for a particle of charge q and mass m under the influence of a magnetic field $\vec{B}$ derived from a vector potential $\vec{A}(\vec{r})$ ($\vec{B} = \vec{\nabla} \times \vec{A}$):

$$
\begin{aligned}
\vec{J}_{em} &= \frac{q}{2m}\left[\psi^*(\vec{r})\left(-i\hbar\vec{\nabla} - q\vec{A}(\vec{r})\right)\psi + \psi(\vec{r})\left(i\hbar\vec{\nabla} - q\vec{A}(\vec{r})\right)\psi^*\right] \\
&= \frac{\hbar q}{m}\left(\vec{\nabla}\theta(\vec{r}) - \frac{q}{\hbar}\vec{A}(\vec{r})\right)\rho(\vec{r}).
\end{aligned}
\tag{6.43}
$$

This expression can be used to show that the magnetic field cannot penetrate the bulk of the superconductor, but must decrease exponentially from its surface (Exercise 6.5.5). Another crucial consequence of (6.43) is that the magnetic flux φ through a superconductor ring is quantized due to the fact that currents must flow at the surface of the superconductor (Exercise 6.5.5). Indeed, considering a contour drawn inside the ring along which the current density vanishes, from $\vec{J}_{em} = 0$ and from the fact that the wave function is single valued we find

$$\oint \vec{\nabla}\theta \cdot \mathrm{d}l = \frac{q_C}{\hbar}\oint \vec{A}\cdot\mathrm{d}\vec{l} = \frac{q_C}{\hbar}\int\int \vec{B}\cdot\mathrm{d}\vec{S} = 2\pi n, \qquad n \in \mathbb{Z},$$

so that

$$\int\int \vec{B}\cdot\mathrm{d}\vec{S} = \frac{\hbar}{q_C}2\pi n = n\varphi_0 \tag{6.44}$$

where q_C is the charge of the Cooper pair, $q_C = 2q_e$, q_e being the electron charge. The elementary *flux quantum* $\varphi_0 = 2\pi\hbar/q_C$, $|\varphi_0| \simeq 2 \times 10^{-15}$ Wb, plays an important role in what follows.

A Josephson junction (Fig. 6.6) is made of two superconductors which are separated by a layer of insulating material. If the layer is thick the electrons cannot cross it, but if it is thin enough they can get across owing to a quantum process known as tunneling. If ψ_1 (ψ_2) is the macroscopic wave function (6.42) on the right (left) side of the junction, to which a bias voltage V is applied, we

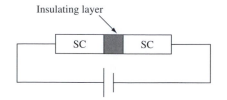

Figure 6.6 Schematic depiction of a Josephson junction: a thin insulating layer is sandwiched between two superconducting (SC) wires.

can describe the junction as a two-level system, and by analogy with (3.30) we can write down a system of coupled differential equations for ψ_1 and ψ_2:

$$
\begin{aligned}
i\hbar\frac{d\psi_1}{dt} &= \frac{1}{2}\,q_C V\psi_1 + K\psi_2, \\
i\hbar\frac{d\psi_2}{dt} &= -\frac{1}{2}\,q_C V\psi_2 + K\psi_1.
\end{aligned}
\tag{6.45}
$$

The coupling between the two wave functions is due to tunneling and is characterized by an amplitude K which may be chosen to be real. The factors $\pm q_C V/2$ can be understood as follows. In the absence of coupling ($K = 0$) the functions ψ_1 and ψ_2 are energy eigenstates whose time evolution is $\exp(\pm iq_C Vt/2\hbar)$ since the energy of a Cooper pair is $\pm q_C V/2$. Writing ψ_1 and ψ_2 in the form (6.42)

$$
\psi_1 = \sqrt{\rho_1}\,e^{i\theta_1}, \qquad \psi_2 = \sqrt{\rho_2}\,e^{i\theta_2},
\tag{6.46}
$$

we obtain a system of coupled equations for the quantities ρ_1, ρ_2, θ_1, and θ_2. After some simple algebra (see Exercise 6.5.6), we find that the Josephson current I_J across the junction is

$$
I_J = \frac{2K}{\hbar}\,\sqrt{\rho_1\rho_2}\,\sin\theta = I_0\sin\theta,
\tag{6.47}
$$

where $\theta = \theta_2 - \theta_1$ is the phase difference across the junction. I_0 is called the critical current, and it is a parameter characteristic of the junction. The phase difference is governed by the equation (Exercise 6.5.6)

$$
V = \frac{\hbar}{q_C}\frac{d\theta}{dt} = \frac{\varphi_0}{2\pi}\frac{d\theta}{dt}.
\tag{6.48}
$$

Note that in a stationary regime the potential difference across the junction vanishes as long as there is no dissipation in the junction. Equations (6.47) and (6.48) are the fundamental equations of the Josephson effect. These two equations can be combined to obtain

$$
\frac{dI_J}{dt} = \frac{2\pi}{\varphi_0}\,VI_0\cos\theta.
\tag{6.49}
$$

This expression can be interpreted as that of an inductor, $\mathcal{E} = L\,dI/dt$, where $\mathcal{E}$ is the emf and L is the inductance, with the effective inductance of the junction given by

$$L_{\mathrm{J}} = \frac{\varphi_0}{2\pi I_0 V \cos\theta}. \tag{6.50}$$

The $1/\cos\theta$ behavior makes it clear that this inductance is highly nonlinear, a property we were looking for. The energy U_{J} stored in the inductor is

$$U_{\mathrm{J}} = \int_{-\infty}^{t} I_{\mathrm{J}}(t')V_{\mathrm{J}}(t')\,dt' = \frac{I_0\varphi_0}{2\pi} \int_0^{\theta} \sin\theta'\,d\theta' = -\frac{I_0\varphi_0}{2\pi} \cos\theta = E_{\mathrm{J}}\cos\theta. \tag{6.51}$$

Before discussing the construction of circuits for superconducting qubits, let us describe a simple closed circuit (a ring) built with one Josephson junction, an inductor L, and a capacitor C (Fig. 6.7). The total magnetic flux through the circuit is φ, composed of an external flux φ_{ex} and the flux $(\varphi - \varphi_{\mathrm{ex}})$ due to the inductance. From a slight modification of (6.44) (see Exercise 6.5.5) we deduce that

$$\theta = 2\pi\frac{\varphi}{\varphi_0} + 2n\pi, \qquad n \in \mathbb{Z}, \tag{6.52}$$

so that $\cos\theta$ in (6.51) can be replaced by $\cos(2\pi\varphi/\varphi_0)$. Then the total energy of the circuit, or the Hamiltonian, is

$$H = \frac{1}{2C}q^2 + \frac{1}{2L}(\varphi - \varphi_{\mathrm{ex}})^2 - \frac{I_0\varphi_0}{2\pi}\cos\left(\frac{2\pi\varphi}{\varphi_0}\right) - I_{\mathrm{ex}}\varphi, \tag{6.53}$$

where we have added a possible contribution from an external current I_{ex} arriving at the ring.

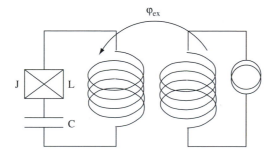

Figure 6.7 A superconducting circuit for flux qubits. J represents the Josephson junction.

As in the case of the *LC* circuit of Fig. 6.5, we can quantize the Hamiltonian H in (6.53) by using two conjugate operators Q and Φ obeying the commutation relation $[Q, \Phi] = i\hbar I$. We then write the quantum Hamiltonian $\hat{H}$ as

$$\hat{H} = \frac{1}{2C} Q^2 + \frac{1}{2L} (\Phi - \varphi_{\text{ex}})^2 - \frac{I_0 \varphi_0}{2\pi} \cos\left(\frac{2\pi\Phi}{\varphi_0}\right) - I_{\text{ex}}\Phi. \qquad (6.54)$$

A possible realization of the canonical commutation relations is $\Phi \to \varphi$, $Q \to i\hbar\partial/\partial\varphi$, where the operators Q and Φ act in the Hilbert space of square integrable functions of φ. In this realization of the commutation relations Φ is represented by multiplication by φ, while Q is represented by the differential operator $i\hbar\partial/\partial\varphi$. Indeed, it can immediately be checked that $[Q, \Phi]$ acting on a function $f(\varphi)$ gives back $i\hbar f(\varphi)$:

$$[Q, \Phi]f(\varphi) = i\hbar\frac{\partial(\varphi f)}{\partial\varphi} - i\hbar\varphi\frac{\partial f}{\partial\varphi} = i\hbar f(\varphi).$$

The circuit of Fig. 6.7 has been used to explore the boundary between the classical and quantum worlds, as it allows the study of quantum superpositions of macroscopic distinguishable states (see Leggett (2002)). The macroscopic distinct states correspond to currents of order $1\,\mu\text{A}$ flowing in opposite directions in the ring.

Three different types of circuit with Josephson junctions have been proposed to serve as a support for qubits. At present it is quite impossible to tell which of these circuits (if any!) will turn out to be the most suitable one. We shall limit our discussion to the "flux qubits," leaving the "charge qubits" to Exercise 6.5.7 and the "phase qubits" to the references. The circuit for flux qubits is that of Fig. 6.7, where an external coil is used to generate an external flux φ_{ex}. The quantum Hamiltonian acting on functions $f(\varphi)$ is then

$$\begin{aligned}
\hat{H} &= \frac{1}{2C}\left(-\hbar^2\frac{\partial^2}{\partial\varphi^2}\right) + \frac{1}{2L}(\varphi - \varphi_{\text{ex}})^2 + E_{\text{J}}\cos\left(\frac{2\pi\varphi}{\varphi_0}\right) \\
&= \frac{1}{2C}\left(-\hbar^2\frac{\partial^2}{\partial\varphi^2}\right) + U(\varphi),
\end{aligned} \qquad (6.55)$$

with $E_{\text{J}} = -I_0\varphi_0/(2\pi)$ (see (6.51)). The flux φ can take continuous values between $-\infty$ and $+\infty$. The Hamiltonian (6.55) contains three adjustable energy scales: E_{J}, $q_{\text{C}}^2/2C$, and $\varphi^2_0/2L$. The flux qubits correspond to the case $E_{\text{J}} \gg q_{\text{C}}^2/2C$, where the phase is well defined and the number of Cooper pairs fluctuates strongly, while the opposite situation holds for the charge qubits. For $\varphi_{\text{ex}} = \varphi_0/2$ it is clear that the potential $U(\varphi)$ in (6.55) is symmetric in the variable $(\varphi - \varphi_0/2)$, and for a suitable choice of the parameters it exhibits a double-well structure (Fig. 6.8). Owing to the symmetry the ground states of the two wells are degenerate. They correspond physically to a macroscopic ($\sim 1\,\mu\text{A}$) current flowing around the loop in the clockwise direction (the state $|0\rangle$) and in the counter-clockwise direction (the state $|1\rangle$). These states $|0\rangle$ and $|1\rangle$ will be chosen as the two basis states

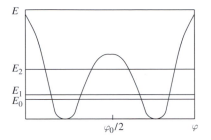

Figure 6.8 The potential $U(\varphi)$ in the symmetric case $\varphi_{\mathrm{ex}} = \varphi_0/2$, together with its three lowest levels.

for a qubit. The symmetric double-well potential $U(\varphi)$ in (6.55) is sketched in Fig. 6.8 together with its first three energy levels for $\varphi_{\mathrm{ex}} = \varphi_0/2$. The two wells are separated by a barrier, and in classical physics a system with low enough energy would be trapped in one of the two wells. This is not the case in quantum physics, where the system can tunnel from one well to the other. Let c_0 and c_1 be the probability amplitudes for finding the system in the state $|0\rangle$ or $|1\rangle$. In the absence of tunneling, the Schrödinger equation would be simply

$$i\hbar \dot{c}_0 = E_0 c_0, \qquad i\hbar \dot{c}_1 = E_0 c_1,$$

where the dynamics of the two states are decoupled and they both have energy E_0. In the presence of tunneling, the two amplitudes will be coupled as

$$i\hbar \frac{dc_0}{dt} = E_0 c_0 - A c_1,$$

$$i\hbar \frac{dc_1}{dt} = E_0 c_1 - A c_0, \tag{6.56}$$

where $A > 0$ is the tunneling amplitude. The two possible values of the energy are $E_0 \mp A$, respectively corresponding to the states $|\pm\rangle$:

$$|\pm\rangle = \frac{1}{\sqrt{2}} (|0\rangle \pm |1\rangle). \tag{6.57}$$

When we move from $\varphi_{\mathrm{ex}} = \varphi_0/2$ to other values of φ_{ex} the symmetry of $U(\varphi)$ about the point $\varphi = \varphi_0/2$ is lost, and one of the wells becomes deeper than the other. The splitting $2E_{\mathrm{ex}}$ between the bottom of the two wells varies linearly with the applied flux

$$E_{\mathrm{ex}} = \zeta \frac{\varphi_0^2}{2L} \left(\frac{\varphi_{\mathrm{ex}}}{\varphi_0} - \frac{1}{2} \right), \tag{6.58}$$

where ζ is a number which must be computed numerically. In the $|\pm\rangle$ basis, the qubit Hamiltonian may be written in the form

$$\hat{H}_{\mathrm{qubit}} = -A\sigma_z + E_{\mathrm{ex}}\sigma_x,$$

the eigenvalues of the energies being

$$E_\pm = \pm\sqrt{A^2 + E_{ex}^2}.$$

For $E_{ex} = 0$ ($\varphi_{ex} = \varphi_0/2$), $E_\pm = \pm A$, while for $|E_{ex}| \gg A$, one has $E_\pm \simeq \pm|E_{ex}|$. The level scheme is drawn in Fig. 6.9, where the familiar level repulsion is observed. Instead of crossing at $\varphi_{ex} = \varphi_0/2$, the two levels "avoid" each other (Exercise 6.5.7). The final result for the qubit Hamiltonian can be written in the form

$$\hat{H}_{qubit} = -A(\sigma_z + X\sigma_x),$$

$$X = -\frac{E_{ex}}{A} = -\frac{\zeta}{A}\frac{\varphi_0^2}{2L}\left(\frac{\varphi_{ex}}{\varphi_0} - \frac{1}{2}\right). \tag{6.59}$$

In practice, a circuit with three Josephson junctions instead of one is used in the case of Fig. 6.7. This allows the behavior of the circuit to be fine tuned more easily. It has been possible to demonstrate Rabi oscillations on the flux qubit as well as Ramsey fringes and spin echo, and to perform spectroscopic measurements on two coupled qubits. The readout of the qubits in the case of superconducting qubits is a delicate problem, because the readout device is a source of decoherence. In the flux qubit case, the readout is performed by means of a SQUID which encircles the circuit. A SQUID, which is also constructed from Josephson junctions, is a very sensitive magnetometer, and measurement of the direction of the magnetic field, which is linked to the sense of rotation of the current, allows the qubit state to be deduced. In recent experiments it has been possible to obtain decoherence times $\sim 1\,\mu$s, whereas a gate operation takes a few nanoseconds. Recent experiments have also shown that it is possible to

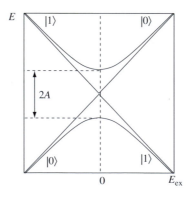

Figure 6.9 The level scheme of the qubit Hamiltonian (6.59). The level repulsion is clearly seen: instead of crossing at $E_{ex} = 0$, the two levels are separated by $2A$. For $|E_{ex}| \gg A$, the eigenstates of $\hat{H}_{qubit}$ are approximately $|0\rangle$ and $|1\rangle$.

couple two superconducting qubits by using a mutual inductance (or capacitance) coupling the two circuits.

6.4 Quantum dots

The amazing progress in the fabrication of small artificial structures in semiconductors has led to an abundance of proposals for solid-state implementations of quantum computers. However, so far none of these proposals has been able to exhibit the concrete realization of a pair of qubits with controlled entanglement together with efficient readout of the final state.

Here we shall consider only one of the most promising schemes, which is based on *quantum dots*. A quantum dot is a structure built in a semiconductor which is able to confine electrons in three dimensions in such a way that discrete energy levels are obtained, much as for an atom; in fact, quantum dots can be regarded as artificial atoms whose characteristics can be controlled by hand. Furthermore, the number of charge carriers in the conduction band of the dot can be very precisely controlled. Quantum dots form spontaneously when semiconductor material is deposited on a substrate with a different lattice spacing. Such a combination is called a heterostructure. The quantum dots have a bowl-like shape, with a typical diameter of 100 nm and a height of 30 nm. The confining potential for charge carriers is approximately a two-dimensional harmonic potential in the horizontal plane.

Two main schemes have been proposed for qubits. The first one is based on excitons, an exciton being an electron–hole pair which is created by light absorption. The exciton energy is $E_{ex} = E_g - E_b$, where E_g is the band gap and E_b is the binding energy of the electron–hole pair. The idea is to use two coupled quantum dots to entangle qubits. The electron and the hole can be in one (the state $|0\rangle$) or the other (the state $|1\rangle$) of the quantum dots. Then $|00\rangle$ corresponds to an electron–hole pair in dot 1, $|11\rangle$ to the pair in dot 2, and $|01\rangle$ and $|10\rangle$ to the electron in dot 1 and the hole in dot 2 and vice versa. If the distance between the two dots is of order 5 nm, electrons and holes can tunnel from one dot to the other. As in the case of superconducting qubits, the eigenstates are therefore the symmetric and antisymmetric linear combinations of the states $|0\rangle$ and $|1\rangle$. The readout of the excitonic states is relatively straightforward, at least in principle. The electron–hole pair recombines after a time ~ 1 ns, and the wavelength of the emitted photon is directly linked to the state occupied by the particles before they recombined.

The second idea is to use electron spin to encode qubits. The first advantage of this scheme is that the Hilbert space for spin is two-dimensional and there is no contamination from other levels. The second one is that decoherence times can be as long as a few microseconds, because the spins are weakly coupled to their environment. The third one is that spins can in principle be transported along conducting wires within the quantum network. Compared to nuclear spins, electron spins have much stronger couplings to magnetic fields, because the ratio of the proton mass to the electron mass is $\sim 10^3$, and the gyromagnetic ratio (Box 3.1) is inversely proportional to the mass. This feature allows gate operations which are much faster than for NMR. The readout of the final state is probably the most difficult challenge for spin qubits. One possibility might be to convert spin degrees of freedom into charge degrees of freedom, followed by electrical detection. Important progress in readout has been recently achieved by Hanson *et al.* (2005).

The Hamiltonian of a set of electron spins localized in a coupled array of quantum dots can be written as

$$\hat{H} = \sum_{\langle i,j \rangle} J_{ij}(t)\vec{\sigma}_i \cdot \vec{\sigma}_j + \frac{1}{2}\mu_B \sum_i g_i(t)\vec{B}_i(t) \cdot \vec{\sigma}_i, \qquad (6.60)$$

where $\vec{\sigma}_i$ is the Pauli spin matrix associated with the ith electron and $\mu_B = q_e\hbar/(2m_e)$ is the Bohr magneton. J_{ij} describes the coupling between spins, which may be assumed to be zero unless the spins are nearest neighbors on the array; $\sum_{\langle ij \rangle}$ denotes a sum over nearest neighbors. The second term in (6.60) is the coupling to the external magnetic field $\vec{B}_i(t)$. As in the NMR case, J_{ij} can be used to build two-qubit gates, while $\vec{B}_i(t)$ is used for single-qubit gates. However, in contrast to the NMR case, the exchange interaction $J_{ij}(t)$ is switched on and off adiabatically. Writing $J_{ij}(t) = J_{ij}[v(t)]$, the necessary condition is $|\dot{v}/v| \ll \delta\varepsilon/\hbar$, where $\delta\varepsilon$ denotes the level separation. In order to implement one-qubit gates, it is necessary to address the right qubit. This can be done by varying the magnetic field $B_i(t)$ or the Landé factor $g_i(t)$ in order to make the resonance frequency dependent on the qubit position. From (6.60) we find the two-qubit Hamiltonian in the form

$$\hat{H} = J(t)\vec{\sigma}_1 \cdot \vec{\sigma}_2. \qquad (6.61)$$

As shown in Exercise 4.6.4, the operator $\vec{\sigma}_1 \cdot \vec{\sigma}_2$ (up to an additive constant) exchanges the two qubits:

$$\vec{\sigma}_1 \cdot \vec{\sigma}_2|00\rangle = |00\rangle, \quad \vec{\sigma}_1 \cdot \vec{\sigma}_2|01\rangle = |10\rangle, \quad \vec{\sigma}_1 \cdot \vec{\sigma}_2|10\rangle = |01\rangle, \quad \vec{\sigma}_1 \cdot \vec{\sigma}_2|11\rangle = |11\rangle.$$

This is the SWAP operation already encountered in (6.37). After a π pulse such that

$$\int_0^{\tau_s} dt J(t) = J_0\tau_s = \pi \bmod 2\pi \qquad (6.62)$$

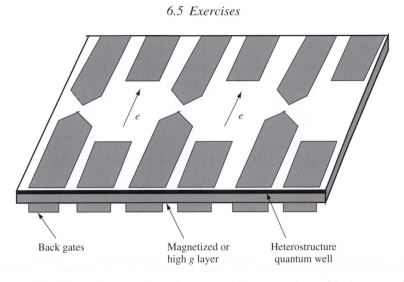

Back gates   Magnetized or   Heterostructure
                    high g layer   quantum well

Figure 6.10 Schematic depiction of an array of quantum dots with electron spins as qubits. After Burkard *et al.* (2002).

we obtain the SWAP gate (6.37). It is easy to pass from a SWAP gate to a cNOT gate owing to the following identity (Exercise 4.6.4). One first obtains a cZ gate

$$cZ = e^{i\pi\sigma_z^1/4}\, e^{-i\pi\sigma_z^2/4}\, U_{\text{SWAP}}^{1/2}\, e^{i\pi\sigma_z^1/2}\, U_{\text{SWAP}}^{1/2}, \tag{6.63}$$

where

$$U_{\text{SWAP}}^{1/2} = \frac{1}{1+i} \begin{pmatrix} 1+i & 0 & 0 & 0 \\ 0 & 1 & i & 0 \\ 0 & i & 1 & 0 \\ 0 & 0 & 0 & 1+i \end{pmatrix},$$

and (6.15) is used for going from a cZ gate to a cNOT gate. The coupling $J(t)$ can be switched on and off by raising and lowering the tunnel barrier between the two dots (Fig. 6.10). In GaAs semiconductors the main source of decoherence comes from the hyperfine coupling to the nuclear spins, as both Ga and As possess a nuclear spin $I = 3/2$. Another source of decoherence is due to the external leads needed for readout.

6.5 Exercises

6.5.1 Off-resonance Rabi oscillations

Starting from the Hamiltonian (6.4) in the rotating reference frame, show that $\exp(-i\tilde{H}t)$ can be written as

$$\exp(-i\tilde{H}t/\hbar) = \exp\left[-\frac{i\Omega t}{2}\left(\frac{\delta}{\Omega}\sigma_z - \frac{\omega_1}{\Omega}\sigma_x\right)\right]$$

with $\Omega = \sqrt{\delta^2 + \omega_1^2}$. The vector

$$\hat{n} = \left(n_x = -\frac{\omega_1}{\Omega}, n_y = 0, n_z = \frac{\delta}{\Omega} \right)$$

is a unit vector. Show that

$$\exp(-i\tilde{H}t/\hbar) = \left(\cos\frac{\Omega t}{2} - i\frac{\delta}{\Omega}\sin\frac{\Omega t}{2} \right) |0\rangle\langle 0| + i\frac{\omega_1}{\Omega}\sin\frac{\Omega t}{2} \left(|0\rangle\langle 1| + |1\rangle\langle 0| \right)$$

$$+ \left(\cos\frac{\Omega t}{2} + i\frac{\delta}{\Omega}\sin\frac{\Omega t}{2} \right) |1\rangle\langle 1|.$$

6.5.2 Commutation relations between the a and a†

1. Use (6.19) and (6.21) to prove the commutation relation $[a, a^\dagger] = I$.

2. Calculate the commutator $[a^\dagger a, a]$. From this derive the second line of (6.30).

6.5.3 Construction of a cZ gate using trapped ions

1. Let us consider the case of a single trapped ion. The laser field applied to the ion has the form (6.27) for $t > 0$. We work in the reference frame rotating at frequency ω_0 (and not ω as in the case of NMR), where the interaction Hamiltonian $\tilde{H}_{\text{int}}(t)$ is given by

$$\tilde{H}_{\text{int}}(t) = e^{i\hat{H}_0 t/\hbar}\,\hat{H}_{\text{int}}(t)\,e^{i\hat{H}_0 t/\hbar}.$$

Show that the rotating-wave approximation leads to the Hamiltonian

$$\tilde{H}_{\text{int}} \simeq -\frac{\hbar}{2}\omega_1 \left[\sigma_+\,e^{i(\delta t - \phi)}\,e^{-ikz} + \sigma_-\,e^{-i(\delta t - \phi)}\,e^{ikz} \right],$$

where $\delta = \omega - \omega_0$ is the detuning. If $\delta = 0$, this Hamiltonian is independent of time:

$$\tilde{H}_1 \simeq -\frac{\hbar}{2}\omega_1 \left[\sigma_+\,e^{-i\phi}\,e^{-ikz} + \sigma_-\,e^{i\phi}\,e^{ikz} \right].$$

2. Let m and $m + m'$ be two levels of the harmonic oscillator. Show that the Rabi frequency $\omega_1^{m \to m+m'}$ is given by

$$\omega_1^{m \to m+m'} = \omega_1 |\langle m + m'|e^{i\eta(a+a^\dagger)}|m\rangle|,$$

where η is the Lamb–Dicke parameter. In particular, show that in the Lamb–Dicke approximation $\eta \ll 1$, and for $m' = \pm 1$ we have the following for the blue and red side-bands (see Fig. 6.4):

$$\omega_1^{m \to m+1} \simeq \eta\sqrt{m+1}\,\omega_1 = \omega_1^+ \quad \text{(blue)},$$

$$\omega_1^{m \to m-1} \simeq \eta\sqrt{m}\,\omega_1 = \omega_1^- \quad \text{(red)}.$$

Derive the Hamiltonian for the two bands, first for the blue side-band

$$\tilde{H}_{\text{int}}^{+} = \frac{i}{2} \eta \hbar \omega_1 \sqrt{m+1} \left[\sigma_+ a_b \, e^{-i\phi} - \sigma_- a_b^\dagger \, e^{i\phi} \right],$$

and then for the red side-band

$$\tilde{H}_{\text{int}}^{-} = \frac{i}{2} \eta \hbar \omega_1 \sqrt{m} \left[\sigma_+ a_r^\dagger \, e^{-i\phi} - \sigma_- a_r \, e^{i\phi} \right].$$

The operators $a_b \cdots a_r^\dagger$ are defined so as to preserve the norm of the state vectors:

$$a_b = \frac{a}{\sqrt{m+1}}, \qquad a_b^\dagger = \frac{a^\dagger}{\sqrt{m+1}}, \qquad a_r = \frac{a}{\sqrt{m}}, \qquad a_r^\dagger = \frac{a^\dagger}{\sqrt{m}}.$$

We limit ourselves to the case $m = 1$. What are the rotation operators on the two bands $R^{\pm}(\theta, \phi)$, where $\theta = -\omega_1^{\pm} t$? $R^{\pm}(\theta, \phi)$ is a rotation by an angle θ about an axis in the xOy plane making an angle ϕ with the axis Ox and using the blue $(+)$ or red $(-)$ side-band.

3. In addition to the levels $|0, 0\rangle$, $|0, 1\rangle$, $|1, 0\rangle$, and $|1, 1\rangle$ of Fig. 6.3(a), we also use the level $|1, 2\rangle$. Sketch the level scheme and identify the transitions of the blue side-band $|0, 0\rangle \leftrightarrow |1, 1\rangle$ and $|0, 1\rangle \leftrightarrow |1, 2\rangle$. Show that the rotation operator $R_{\alpha\beta}^{+}$, defined as

$$R_{\alpha\beta}^{+} = R^{+}(\alpha, \pi/2) \, R^{+}(\beta, 0) \, R^{+}(\alpha, \pi/2) \, R^{+}(\beta, 0),$$

is equal to $-I$ for $\alpha = \pi$ and any β or for $\beta = \pi$ and any α. Using the fact that the Rabi frequency for the transition $|0, 1\rangle \leftrightarrow |1, 2\rangle$ is $\sqrt{2}$ times that for the transition $|0, 0\rangle \leftrightarrow |1, 1\rangle$, how can α and β be chosen such that $R_{\alpha\beta}^{+} = -I$ for the two transitions? Find the sequence of 4 pulses and their duration such that the net result is

$$|00\rangle \leftrightarrow -|0, 0\rangle, \qquad |0, 1\rangle \leftrightarrow -|0, 1\rangle, \qquad |1, 0\rangle \leftrightarrow +|1, 0\rangle, \qquad |1, 1\rangle \leftrightarrow -|1, 1\rangle.$$

This is how a cZ gate is constructed (up to a sign).

4. Now we need to "transfer" the cZ gate to the computational basis of the states $|n_1, n_2\rangle$, $n_1, n_2 = 0, 1$ being the ground and excited states of the two ions. Show that the desired result is obtained by sandwiching the rotation operator $R_{\alpha\beta}^{+(1)}$ on ion number 1 using the blue side-band between two π rotations on ion number 2 using the red side-band:

$$\left[R^{-(2)}(\pi, \pi/2) \right] R_{\alpha\beta}^{+(1)} \left[R^{-(2)}(-\pi, \pi/2) \right].$$

Using (6.15), one goes from a cZ gate to a cNOT gate, but a slightly more complicated operation allows the direct construction of a cNOT gate.

6.5.4 Vibrational normal modes of two ions in a trap

The potential energy of the two ions is

$$V = \frac{1}{2} M\omega_z^2 \left(z_1^2 + z_2^2\right) + \frac{q^2}{4\pi\varepsilon_0} \frac{1}{|z_1 - z_2|}.$$

Find the equilibrium positions of the two ions. Show that the vibrational eigen-frequencies are ω_z and $\sqrt{3}\omega_z$. How can the vibrational amplitudes of these two normal modes be characterized? Hint: the equilibrium positions being $\pm z_0$, write $z_1 = z_0 + u$, $z_2 = -z_0 + v$, and expand V to second order in (u, v).

6.5.5 Meissner effect and flux quantization

1. Starting from (6.43) and using the Maxwell equation $\vec{\nabla} \times \vec{B} = \mu_0 \vec{j}_{em}$, show that the magnetic field $\vec{B}$ obeys

$$\nabla^2 \vec{B} = \frac{1}{\lambda_L^2} \vec{B},$$

where λ_L is the London penetration length. Give the expression for λ_L. Assume a one-dimensional geometry, where the region $z < 0$ is normal while the region $z > 0$ is superconducting. Show that $\vec{B}$ and, consequently, $\vec{j}_{em}$ decrease as $\exp(-z/\lambda_L)$. The electromagnetic current and the magnetic fields are therefore excluded from the bulk of the superconducting region.

2. Consider a contour C drawn in the bulk of a superconducting ring. Using $\vec{j}_{em} = 0$ along C, derive (6.44) and (6.52).

6.5.6 Josephson current

By separating the real and imaginary parts in the equations (6.45) for ψ_1 and ψ_2, show that

$$\frac{d\rho_1}{dt} = \frac{2K}{\hbar}(\rho_1\rho_2)^{1/2} \sin\theta,$$

$$\frac{d\rho_2}{dt} = -\frac{2K}{\hbar}(\rho_1\rho_2)^{1/2} \sin\theta,$$

$$\frac{d\theta_1}{dt} = -\frac{K}{\hbar}\left(\frac{\rho_2}{\rho_1}\right)^{1/2} \cos\theta - \frac{q_C V}{2\hbar},$$

$$\frac{d\theta_2}{dt} = \frac{K}{\hbar}\left(\frac{\rho_1}{\rho_2}\right)^{1/2} \cos\theta + \frac{q_C V}{2\hbar}.$$

Assuming that $\rho_1 \simeq \rho_2 = \rho$, deduce from these equations the Josephson current I_J:

$$I_J = \frac{d\rho_1}{dt} = -\frac{K\rho}{\hbar} \sin\theta = I_0 \sin\theta$$

and

$$\frac{d\theta}{dt} = \frac{q_C V}{\hbar}.$$

6.5.7 Charge qubits

The circuit is drawn in Fig. 6.11. The superconducting box (Cooper pair box) is divided into two parts by a Josephson junction. If C_J and C_g denote the capacitances of the junction and the external capacitor, respectively, and V_g is the applied voltage bias, the electrical energy stored in the circuit reads

$$E = E_c(n - n_g)^2 - E_J \cos\theta,$$

where θ is the junction phase difference. The energy E_c,

$$E_c = \frac{q_C^2}{2(C_J + C_g)},$$

is the electrostatic energy of the capacitors and $n (n \in \mathbb{Z})$ is the number of *excess* Cooper pairs in the right-hand box. Let N be the operator which counts the number of *excess* Cooper pairs in the right-hand box, with eigenstates $|n\rangle$, $N|n\rangle = n|n\rangle$, and let Θ be the conjugate operator, $[N, \Theta] = iI$. The eigenstates of Θ are denoted $|\theta\rangle$, $\Theta|\theta\rangle = \theta|\theta\rangle$, $0 \le \theta < 2\pi$.

1. Show that if we choose the scalar product $\langle n|\theta\rangle = \exp(-in\theta)$ such that

$$|\theta\rangle = \sum_{n=-\infty}^{+\infty} e^{-in\theta} |n\rangle,$$

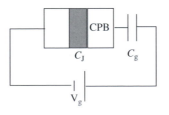

Figure 6.11 Superconducting circuit for charge qubits. CPB, Cooper pair box.

then $N = i\partial/\partial\theta$ and the commutation relation $[N, \Theta] = i$ is satisfied. Note that the bases $\{|n\rangle\}$ and $\{\theta\rangle\}$ are complementary according to the definition of Section 2.4. Derive the completeness relation

$$\int_0^{2\pi} \frac{d\theta}{2\pi} |\theta\rangle\langle\theta| = I$$

and deduce from this relation that

$$\cos\Theta = \frac{1}{2} \sum_{n=-\infty}^{+\infty} (|n\rangle\langle n+1| + |n+1\rangle\langle n|).$$

2. In the $\{|n\rangle\}$ basis the quantum Hamiltonian reads

$$\hat{H} = E_c \sum_{n=-\infty}^{+\infty} (n - n_g)^2 |n\rangle\langle n| - \frac{1}{2} E_J \sum_{n=-\infty}^{+\infty} (|n\rangle\langle n+1| + |n+1\rangle\langle n|).$$

We assume that $E_J \ll E_c$, so that the first term in $\hat{H}$ is dominant except when $n_g \simeq p + 1/2$ with p an integer, in which case the first term gives the same energy $E_c/4$ for the states $n = p$ and $n = p+1$. In the absence of the second term the two states would be exactly degenerate for $n_g = p + 1/2$. For definiteness let us choose $n_g = 1/2$. Show that in the vicinity of $n_g = 1/2$ we can write

$$\hat{H} = -\frac{1}{2} B_z \sigma_z - \frac{1}{2} B_x \sigma_x,$$

where $B_x = 0$ if $n_g = 1/2$. Determine B_z and B_x as functions of the parameters in $\hat{H}$. Draw the level scheme at fixed B_x as a function of B_z, discuss the phenomenon of level repulsion, and show that the level scheme is similar to that of Fig. 6.9.

6.6 Further reading

The criteria for a quantum computer have been stated by di Vincenzo (2000). The concrete realizations of quantum computers are described by Nielsen and Chuang (2000), Chapter 7, by Bouwmeester *et al.* (2000) and by Stolze and Suter (2004), Chapters 9 to 12. The experimentally performed (NMR) factorization of 15 using the Shor algorithm was realized by Vandersypen *et al.* (2001); see also Vandersypen and Chuang (2004). The article by Cirac and Zoller (2004) describes recent results on trapped ions and Bose–Einstein condensates, and that of Mooij (2004) describes the results obtained using Josephson junctions; see also You and Nori (2005). Another useful review on trapped ions is Leibfried *et al.* (2003). The standard results on the quantum harmonic oscillator can be found in any quantum mechanics textbook, for example, Le Bellac (2006), Chapter 11; Doppler cooling is described in Section 14.4 of the same book. A complete panorama of the recent developments in quantum information theory is given at an advanced level in the proceedings of the Les Houches School 2003; see, in particular, the courses of Jones (2004), Blatt (2004), and Devoret and Martinis (2004). The use of quantum dots is described by Burkard *et al.* (2002).

7

Quantum information

In this chapter we address the problems of the storage and transmission of quantum information. Quantum cryptography should in principle be included here, but we preferred to describe it in Chapter 2, where it was used to illustrate the basic principles of quantum mechanics. In Section 7.1 we explain quantum teleportation and in Section 7.2 we give a short and schematic review of classical information theory. Section 7.3 is devoted to an introduction to the storage and communication of quantum information, and, finally, in Section 7.4 we take a quick look at the important but difficult topic of quantum error correction.

7.1 Teleportation

Teleportation is an interesting application of entangled states which may have applications to quantum information transfer (Fig. 7.1). Let us suppose that Alice wishes to transfer to Bob the information about the spin state $|\varphi_A\rangle$ of a particle A of spin 1/2,

$$|\varphi_A\rangle = \lambda|0_A\rangle + \mu|1_A\rangle, \tag{7.1}$$

which is *a priori* unknown, without sending Bob this particle directly. She cannot measure its spin, because she does not know the basis in which the spin of particle A was prepared, and any measurement would in general project $|\varphi_A\rangle$ onto another state. The principle of information transfer consists of using an auxiliary pair of entangled particles B and C of spin 1/2 shared between Alice and Bob. Particle B is used by Alice and particle C is sent to Bob (Fig. 7.1). These particles B and C may be, for example, in an entangled spin state $|\Psi_{BC}\rangle$:

$$|\Psi_{BC}\rangle = \frac{1}{\sqrt{2}}\left(|0_B 0_C\rangle + |1_B 1_C\rangle\right). \tag{7.2}$$

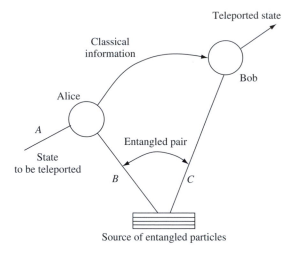

Figure 7.1 Teleportation. Alice makes a Bell measurement on the qubits A and B and informs Bob of the result by a classical path.

The initial state of the three particles $|\Phi_{ABC}\rangle$ then is

$$|\Phi_{ABC}\rangle = (\lambda|0_A\rangle + \mu|1_A\rangle)\, \frac{1}{\sqrt{2}}\, (|0_B0_C\rangle + |1_B1_C\rangle)$$

$$= \frac{\lambda}{\sqrt{2}}|0_A\rangle\, (|0_B0_C\rangle + |1_B1_C\rangle) + \frac{\mu}{\sqrt{2}}|1_A\rangle\, (|0_B0_C\rangle + |1_B1_C\rangle)\,. \tag{7.3}$$

Alice first applies a cNOT gate to the qubits A and B, with the qubit A acting as the control qubit and the qubit B acting as the target qubit (Fig. 7.2). This operation transforms the initial state (7.3) of three qubits into

$$|\Phi'_{ABC}\rangle = \frac{\lambda}{\sqrt{2}}\, (|0_A\rangle(|0_B0_C\rangle + |1_B1_C\rangle) + \frac{\mu}{\sqrt{2}}\, (|1_A\rangle(|1_B0_C\rangle + |0_B1_C\rangle)\,. \tag{7.4}$$

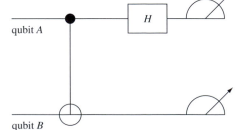

Figure 7.2 Alice applies a cNOT gate to the qubits A and B and then a Hadamard gate to the qubit A.

Alice then applies a Hadamard gate to the qubit A, which transforms (7.4) into

$$|\Phi''_{ABC}\rangle = \frac{1}{2}\Big[\lambda|0_A 0_B 0_C\rangle + \lambda|0_A 1_B 1_C\rangle + \lambda|1_A 0_B 0_C\rangle + \lambda|1_A 1_B 1_C\rangle$$

$$+\mu|0_A 1_B 0_C\rangle + \mu|0_A 0_B 1_C\rangle - \mu|1_A 1_B 0_C\rangle - \mu|1_A 0_B 1_C\rangle\Big]. \tag{7.5}$$

This equation can be rewritten as

$$|\Phi''_{ABC}\rangle = \frac{1}{2}|0_A 0_B\rangle\,(\lambda|0_C\rangle + \mu|1_C\rangle)$$

$$+\frac{1}{2}|0_A 1_B\rangle\,(\mu|0_C\rangle + \lambda|1_C\rangle)$$

$$+\frac{1}{2}|1_A 0_B\rangle\,(\lambda|0_C\rangle - \mu|1_C\rangle) \tag{7.6}$$

$$+\frac{1}{2}|1_A 1_B\rangle\,(-\mu|0_C\rangle + \lambda|1_C\rangle).$$

The last operation that Alice performs is measurement of the two qubits in the basis $\{|0\rangle, |1\rangle\}$. The joint measurement by Alice of qubits A and B is called a *Bell measurement*. This measurement projects the pair (AB) onto one of the four states $|i_A j_B\rangle$, $i, j = 0, 1$, and the state vector of the qubit C is then read on each of the lines of (7.6).

The simplest case occurs when the measurement result is $|0_A 0_B\rangle$. The qubit C then reaches Bob in the state

$$\lambda|0_C\rangle + \mu|1_C\rangle,$$

that is, in the initial state of the qubit A, with the *same* coefficients λ and μ. So, Alice informs Bob by a classical channel (for example, a telephone) that the qubit will reach him in the same state as the qubit A. If on the contrary she measures $|0_A 1_B\rangle$, the qubit C is in the state

$$\mu|0_C\rangle + \lambda|1_C\rangle,$$

she then informs Bob that he must apply to qubit C a rotation of π about Ox, or, equivalently, the matrix σ_x:

$$\exp\left(-i\frac{\pi\sigma_x}{2}\right) = -i\sigma_x.$$

In the third case ($|1_A 0_B\rangle$) it is necessary to apply a rotation of π about Oz, and in the final case ($|1_A 1_B\rangle$) a rotation of π about Oy. We note that in these four cases Alice does not know the coefficients λ and μ, and she sends Bob only the information about which rotation he should apply.

It is useful to add a few final remarks.

- The coefficients λ and μ are never measured, and the state $|\varphi_A\rangle$ is destroyed during Alice's measurement. There is therefore no contradiction with the no-cloning theorem.
- Bob "knows" the state of particle C only once he has received the result of Alice's measurement. This information must be sent by a classical channel, at a speed at most equal to that of light. There is therefore no instantaneous transmission of information at a distance.
- Teleportation never involves the transport of matter.

7.2 Shannon entropy

The two fundamental theorems of information theory were stated by Shannon in 1948. Before discussing their quantum generalization, let us give a very schematic review of these theorems without going into detailed proofs. These theorems answer the following questions.

(i) What is the maximal compression that can be applied to a message? In other words, how can redundant information be quantified?
(ii) At what rate can one communicate via a noisy channel, that is, what redundancy must be incorporated in a message to protect against errors?

It can easily be seen from the following example that a message can be compressed when compared to its naive encoding. Let us suppose that we are using four different letters, (a_0, a_1, a_2, a_3), which we can encode in the usual manner using two bits: $a_0 = 00$, $a_1 = 01$, $a_2 = 10$, and $a_3 = 11$. A message n letters long will then be encoded by $2n$ bits. However, suppose that the letters occur with different probabilities: a_0 with probability 1/2, a_1 with probability 1/4, and a_2 and a_3 with probability 1/8. We can then use the following encoding: $a_0 = 0$, $a_1 = 10$, $a_2 = 110$, and $a_3 = 111$. This can easily be verified to be unambiguous: a letter stops after every 0, or after a sequence of three 1. The average length of a message n letters long will then be

$$n\left(\frac{1}{2}1 + \frac{1}{4}2 + \frac{1}{4}3\right) = \frac{7}{4}n < 2n.$$

Shannon's first theorem shows that this is in fact the best possible compression. Let us take a set of letters a_x, $0 \le x \le k$, and a sequence $\{a_1, \ldots, a_n\}$ of n letters forming a message. Each letter occurs *a priori* with a probability $p(a_x)$, $\sum_x p(a_x) = 1$. We consider a message n letters long, $n \gg 1$. Is it possible to compress the message into a shorter sequence containing essentially the same information? The simplest case is that of two letters, $p(a_0) = p$, $p(a_1) = 1 - p$. The probability $p(q)$ that an n-letter message contains q letters a_0 is [1]

$$p(q) = \binom{n}{q} p^q (1-p)^{n-q}.$$

Let us find the maximum of $p(q)$ by using Stirling's formula $\ln n! \simeq n \ln n/e$, from which $d \ln n!/dn \simeq \ln n$. We compute the q-derivative of $p(q)$

$$\frac{d \ln p(q)}{dq} = -\ln q + \ln(n-q) + \ln p - \ln(1-p)$$

which vanishes for $q = \bar{q} = np$. As was to be expected, the most probable value [2] of q is $\bar{q} = np$. The dispersion around $\bar{q}$ is found from the second derivative of $\ln p(q)$

$$\frac{d^2 \ln p(q)}{dq^2}\bigg|_{q=\bar{q}} = -\frac{1}{np(1-p)}$$

so that

$$\langle \Delta q^2 \rangle = \langle (q - \bar{q})^2 \rangle = np(1-p).$$

With a negligible error when $n \to \infty$, the variable q lies in the range

$$np - \mathcal{O}(\sqrt{n}) \lesssim q \lesssim np + \mathcal{O}(\sqrt{n}).$$

The number of occurrences of the letter a_0 in an n-letter message will lie in this range, and the number of typical messages (or sequences) will be of order $\binom{n}{np}$. Instead of coding 2^n sequences, it is sufficient to code the $\simeq \binom{n}{np}$ *typical sequences*. Stirling's formula allows us to compute $\ln \binom{n}{np}$

$$\ln \binom{n}{np} \simeq -n[p \ln p + (1-p) \ln(1-p)] = n\overline{H}_{\mathrm{Sh}}(p),$$

or

$$\binom{n}{np} \simeq e^{n\overline{H}_{\mathrm{Sh}}(p)} = 2^{nH_{\mathrm{Sh}}(p)}.$$

In information theory it is usual to work with base-2 logarithms, and the *Shannon entropy* is then defined by the second expression in the preceding equation, or

$$H_{\mathrm{Sh}}(p) = -p \log p - (1-p) \log(1-p), \tag{7.7}$$

---

[1] We assume that the correlations between letters can be neglected.
[2] The function $p(q)$ is approximately Gaussian around $q = \bar{q}$, so that the most probable value is also the mean value $\langle q \rangle$.

where log is a base-2 logarithm. The number of *typical* sequences is of order $2^{nH_{Sh}(p)}$. Let us illustrate this by two limiting cases.

(i) $p = 1$. In this case the 2^n messages are identical and it is sufficient to send only one: $H_{Sh}(p) = 0$.

(ii) $p = 1/2$. All messages are equally probable and $H_{Sh} = 1$. In this case it is necessary to send the 2^n messages and no compression is possible.

In an intermediate case, for example, $p = 1/4$, it is sufficient to encode typical sequences, and one never has to encode sequences of letters containing very few a_0 or very few a_1, which are highly unlikely.

In the case of k letters a_x with probabilities $p(x)$, the number of typical sequences is

$$\frac{n!}{\prod_x (np(x))!} \simeq 2^{nH_{Sh}(X)}$$

with

$$\boxed{H_{Sh}(X) = - \sum_{x=0}^{k} p(x) \log p(x)} \tag{7.8}$$

where X denotes the probability distribution of the a_x. It can be rigorously shown that if $n \to \infty$, an optimal encoding compresses each letter into $H_{Sh}(X)$ bits. This is the content of the first Shannon theorem, which also states that no further data compression is possible without introducing errors. In the example given above

$$-\left(\frac{1}{2} \log \frac{1}{2} + \frac{1}{4} \log \frac{1}{4} + \frac{1}{4} \log \frac{1}{8} \right) = \frac{7}{4},$$

which shows that the proposed encoding is optimal.

Let us now turn to the problem of a noisy channel. Let $p(y|x)$ be the *conditional* probability for y to be read when the letter[3] x is sent, the letter x being sent with probability $p(x)$. The entropy H_{Sh} quantifies our *a priori* ignorance per letter before receiving the message. Once y is known, we have at our disposal supplementary information, and our ignorance is not as great. We shall make use of Bayes' law

$$p(x|y) = \frac{p(x, y)}{p(y)} \implies p(x|y) = \frac{p(y|x)p(x)}{p(y)} \tag{7.9}$$

and of

$$p(y) = \sum_x p(y|x)p(x).$$

---

[3] To simplify the notation we write $a_x = x$.

The number of bits needed to send a message knowing that y is read is then

$$H_{\text{Sh}}(X|Y) = \langle -\log p(x|y) \rangle = \sum_y p(y) \sum_x p(x|y) \ln p(x|y)$$

$$= H_{\text{Sh}}(X, Y) - H_{\text{Sh}}(Y). \tag{7.10}$$

The *information gain* or *mutual information* $I(X:Y)$ quantifies the information that is acquired about x when y is read:

$$\boxed{I(X:Y) = I(Y:X) = H_{\text{Sh}}(X) - H_{\text{Sh}}(X|Y) = H_{\text{Sh}}(X) + H_{\text{Sh}}(Y) - H_{\text{Sh}}(X, Y)} \tag{7.11}$$

In other words, $I(X:Y) = I(Y:X)$ is the number of bits per letter of X which can be acquired by reading Y (or vice versa). If $p(y|x)$ characterizes a noisy channel, $I(X:Y)$ is the information per letter which can be sent via the channel given the probability distribution X, and the *transmission capacity* C of the channel is the maximum of $I(X:Y)$ over the ensemble of these probability distributions:

$$\boxed{C = \max_{\{p(x)\}} I(X:Y)} \tag{7.12}$$

The second Shannon theorem states that error-free transmission by a noisy channel is possible if the transmission rate of the channel is less than C.

Let us give an example for a symmetric binary channel, defined as

$$p(x = 0|y = 0) = p(x = 1|y = 1) = 1 - p,$$

$$p(x = 0|y = 1) = p(x = 1|y = 0) = p,$$

where the mutual information is

$$I(X:Y) = H_{\text{Sh}}(X) - H_{\text{Sh}}(p),$$

$H_{\text{Sh}}(X)$ being given by (7.8). The maximal value of $H_{\text{Sh}}(X)$ is 1, and so

$$C(p) = 1 - H_{\text{Sh}}(p).$$

Another illustration of the concept of information gain is given in Exercise 7.5.3, where it is applied to quantum cryptography.

7.3 von Neumann entropy

In the quantum case, the letters are replaced by quantum states $|\alpha\rangle$ whose frequency is p_α. The state operator is

$$\rho = \sum_\alpha p_\alpha |\alpha\rangle\langle\alpha|, \quad \sum_\alpha p_\alpha = 1. \tag{7.13}$$

The state operator ρ represents a statistical mixture of states $|\alpha\rangle$, each state $|\alpha\rangle$ having probability p_α. The states $|\alpha\rangle$ are normalized ($\langle\alpha|\alpha\rangle = 1$) but *not necessarily orthogonal* ($\langle\alpha|\beta\rangle \neq \delta_{\alpha\beta}$), and in general there exist an infinite number of decompositions of ρ of the type (7.13). It can also be said that there are an infinite number of ways of preparing ρ. The way it is prepared determines ρ, but not the reverse. However, ρ is Hermitian and can always be diagonalized:

$$\rho = \sum_i \mathsf{p}_i |i\rangle\langle i|, \quad \langle i|j\rangle = \delta_{ij}. \tag{7.14}$$

This leads to a generalization of the Shannon entropy, the *von Neumann entropy*, which is independent of the preparation:

$$\boxed{H_{\mathrm{vN}} = -\sum_i \mathsf{p}_i \log \mathsf{p}_i = -\mathrm{Tr}\rho\log\rho} \tag{7.15}$$

We note that the entropy of a pure case is zero, because all the p_i are zero except for one, which is unity. As in the classical case, we define the Shannon entropy of the preparation (7.13) as

$$H_{\mathrm{Sh}} = -\sum_\alpha \mathsf{p}_\alpha \log \mathsf{p}_\alpha. \tag{7.16}$$

As already mentioned, there are in general an infinite number of different statistical mixtures $\{\mathsf{p}_\alpha, |\alpha\rangle\}$ which give the same state operator, and it can be shown that the Shannon entropy is always greater than the von Neumann entropy (see Exercise 7.5.2):

$$-\sum_\alpha \mathsf{p}_\alpha \log \mathsf{p}_\alpha \geq -\mathrm{Tr}\,\rho\log\rho, \quad H_{\mathrm{Sh}} \geq H_{\mathrm{vN}}. \tag{7.17}$$

As we shall see later on, the entropy H_{vN} quantifies the incompressible information contained in the source described by the state operator ρ. The difference between the Shannon entropy and the von Neumann entropy is particularly clear for a product state AB represented by a state operator ρ_{AB}. The operator ρ_{AB} is used to construct the state operators of A and B, ρ_A and ρ_B, by taking the trace of ρ_{AB} over the spaces $\mathcal{H}_B$ and $\mathcal{H}_A$, respectively [*cf.* (4.12)]:

$$\rho_A = \mathrm{Tr}_B\rho_{AB}, \quad \rho_B = \mathrm{Tr}_A\rho_{AB},$$

or in matrix form [4]

$$\rho_{ij}^A = \sum_\mu \rho_{i\mu,j\mu}^{AB}, \quad \rho_{\mu\nu}^B = \sum_i \rho_{i\mu,i\nu}^{AB}.$$

The operators ρ_A and ρ_B are the reduced state operators of A and B. The following inequalities can be derived for the von Neumann entropy:

$$|H_{\mathrm{vN}}(\rho_A) - H_{\mathrm{vN}}(\rho_B)| \leq H_{\mathrm{vN}}(\rho_{AB}) \leq H_{\mathrm{vN}}(\rho_A) + H_{\mathrm{vN}}(\rho_B). \tag{7.18}$$

---

[4] Here AB is written as a superscript to make room for the subscripts labeling the matrix elements.

On the contrary, the Shannon entropy of a joint probability distribution $H_{Sh}(p_{AB})$ satisfies

$$\max[H_{Sh}(p_A), H_{Sh}(p_B)] \leq H_{Sh}(p_{AB}) \leq H_{Sh}(p_A) + H_{Sh}(p_B), \qquad (7.19)$$

where p_A and p_B are the probability distributions of x_A and x_B

$$p_A(x_A) = \sum_{x_B} p_{AB}(x_A, x_B), \quad p_B(x_B) = \sum_{x_A} p_{AB}(x_A, x_B).$$

The inequality on the right-hand side is the same for the two entropies, but that on the left (called the Araki–Lieb inequality) is different. For example, if ρ_{AB} is the state operator describing the pure state (4.4) of two qubits we have $H_{vN}(\rho_{AB}) = 0$, whereas

$$H_{vN}(\rho_A) = H_{vN}(\rho_B) = 1.$$

The von Neumann entropy provides the key to the quantum generalization of the two Shannon theorems on data compression and on the maximum transmission capacity of a noisy channel. To explain this, let us consider an ensemble of n letters, where each letter is drawn from an ensemble $\{p_\alpha, |\alpha\rangle\}$ such that the state operator of a single letter is given by (7.13). Successive letters are assumed to be independent, and the state operator of the ensemble of letters is

$$\rho^{\otimes n} = \rho \otimes \rho \otimes \cdots \otimes \rho := \sigma, \quad n \gg 1.$$

Let us suppose that we wish to send (or store) a message of n letters, by trying to encode the quantum system in a smaller system. This smaller system is sent to one end of a channel and decoded at the other end. The state operator of the transmitted system is σ', and the *fidelity* $\mathcal{F}$ of the transmission is defined as [5]

$$\mathcal{F}(\sigma, \sigma') := \left(\text{Tr} \sqrt{\sigma^{1/2} \sigma' \sigma^{1/2}} \right)^2. \qquad (7.20)$$

This expression is not very intuitive and does not even look symmetric in σ and σ', although one may prove that *it is symmetric* (see Exercise 7.5.4)

$$\mathcal{F}(\sigma, \sigma') = \mathcal{F}(\sigma', \sigma).$$

If σ is a pure state, $\sigma = \sigma^{1/2} = |\psi\rangle\langle\psi|$ and σ' a state matrix of the form (7.13)

$$\sigma' = \sum_\tau p'_\tau |\tau\rangle\langle\tau|,$$

---

[5] Many authors, including Nielsen and Chuang (2000), define the fidelity as the square root of $\mathcal{F}$ in (7.20).

then

$$\sigma^{1/2}\sigma'\sigma^{1/2} = \left(\sum_\tau \mathsf{p}'_\tau |\langle\psi|\tau\rangle|^2\right)|\psi\rangle\langle\psi|$$

and one finds

$$\mathcal{F}(|\psi\rangle\langle\psi|, \sigma') = \sum_\tau \mathsf{p}'_\tau |\langle\psi|\tau\rangle|^2 = \langle\psi|\sigma'|\psi\rangle. \tag{7.21}$$

When σ and σ' represent pure states $|\psi\rangle$ and $|\psi'\rangle$, the fidelity reduces to

$$\mathcal{F}\left(|\psi\rangle\langle\psi|, |\psi'\rangle\langle\psi'|\right) = |\langle\psi|\psi'\rangle|^2 = \mathsf{p}(\psi' \to \psi)$$

according to (2.18), which in this case is a natural definition because $\mathcal{F}$ is simply the overlap of the two states.

We wish to find the smallest possible system such that $\mathcal{F} \geq 1 - \varepsilon$, for ε arbitrarily small and when letters are qubits. The Hilbert space $\mathcal{H}^{\otimes n}$ of n qubits has dimension 2^n. However, if $H_{\mathrm{vN}}(\rho) < 1$, we are going to show that the state operator can be restricted to a typical Hilbert subspace of $\mathcal{H}^{\otimes n}$, and this *typical subspace* will have dimension smaller than 2^n. The fundamental result of Shumacher (and Josza) is that the dimension of this subspace is $2^{nH_{\mathrm{vN}}(\rho)}$ for $n \gg 1$. It is therefore sufficient to use $nH_{\mathrm{vN}}(\rho)$ qubits to represent faithfully the quantum information. This result transposes the classical result of Shannon, with the idea of a typical sequence of letters replaced by that of a typical subspace, and the Shannon entropy replaced by the von Neumann entropy.

Before giving the proof of Schumacher's theorem, let us explain intuitively why such a compression of qubits is possible. Suppose Alice has drawn her qubits from the ensemble

$$|0\rangle = \begin{pmatrix} 1 \\ 0 \end{pmatrix}, \qquad \mathsf{p} = \frac{1}{2},$$

$$|+\rangle = \frac{1}{\sqrt{2}} \begin{pmatrix} 1 \\ 1 \end{pmatrix}, \qquad \mathsf{p} = \frac{1}{2}, \tag{7.22}$$

so that the state matrix (7.13) is

$$\rho = \frac{1}{4}\begin{pmatrix} 3 & 1 \\ 1 & 1 \end{pmatrix}. \tag{7.23}$$

Observing that $|0\rangle$ is an eigenstate of σ_z and $|+\rangle$, an eigenstate of σ_x, both with eigenvalue $+1$, it is obvious from symmetry considerations that the eigenstates of ρ are the vectors $|0'\rangle = |0, \hat{n}\rangle$ and $|1'\rangle = |1, \hat{n}\rangle$, where $\hat{n}$ is the unit vector $(\hat{x} + \hat{z})/\sqrt{2}$:

$$|0'\rangle = |0, \hat{n}\rangle = \begin{pmatrix} \cos \pi/8 \\ \sin \pi/8 \end{pmatrix} = \begin{pmatrix} \beta \\ \gamma \end{pmatrix},$$

$$|1'\rangle = |1, \hat{n}\rangle = \begin{pmatrix} -\sin \pi/8 \\ \cos \pi/8 \end{pmatrix} = \begin{pmatrix} -\gamma \\ \beta \end{pmatrix}. \tag{7.24}$$

The vectors $|0, \hat{n}\rangle$ and $|1, \hat{n}\rangle$ are eigenvectors of $\vec{\sigma} \cdot \hat{n}$ with eigenvalues $+1$ and -1, respectively. The eigenvalues of ρ are

$$\lambda(0') = \cos^2 \pi/8 = \beta^2 \simeq 0.8536$$

and

$$\lambda(1') = \sin^2 \pi/8 = \gamma^2 \simeq 0.1464.$$

By construction, the state $|0'\rangle$ has the same (large) overlap with $|0\rangle$ and $|+\rangle$:

$$|\langle 0'|0\rangle|^2 = |\langle 0'|+\rangle|^2 = \beta^2 \simeq 0.8536,$$

while $|1'\rangle$ has the same (small) overlap with $|0\rangle$ and $|+\rangle$:

$$|\langle 1'|0\rangle|^2 = |\langle 1'|+\rangle|^2 = \gamma^2 \simeq 0.1564.$$

If we do not know which of the states (7.22) was sent, our best guess is $|\psi\rangle = |0'\rangle$, and from (7.21) the probability of a successful guess is just the fidelity:

$$\frac{1}{2} \left(|\langle 0'|0\rangle|^2 + |\langle 0'|+\rangle|^2 \right) = \mathcal{F}(\rho, |0'\rangle\langle 0'|) \simeq 0.8536.$$

Another way of obtaining the preceding result is to start from a trial vector

$$|\varphi\rangle = \cos \frac{\theta}{2} |0\rangle + e^{i\phi} \sin \frac{\theta}{2} |1\rangle,$$

compute the fidelity $\mathcal{F}(\rho, |\varphi\rangle\langle\varphi|)$, and then check that it has a maximum for $\theta = \pi/4$ (see Exercise 2.6.4).

Now, suppose that Alice wants to send Bob a three-qubit message compressed into only two qubits with maximum fidelity. Let the message be

$$|\Psi\rangle = |\psi_A \otimes \psi_B \otimes \psi_C\rangle,$$

where $|\psi_i\rangle = |0_i\rangle$ or $|+_i\rangle$, and let us examine the three-qubit Hilbert space $\mathcal{H}^{\otimes 3}$. A possible basis of $\mathcal{H}^{\otimes 3}$ is

$$|b_1\rangle = |0'_A 0'_B 0'_C\rangle, \quad |b_2\rangle = |0'_A 0'_B 1'_C\rangle, \quad |b_3\rangle = |0'_A 1'_B 0'_C\rangle, \quad |b_4\rangle = |1'_A 0'_B 0'_C\rangle,$$

$$|b_5\rangle = |0'_A 1'_B 1'_C\rangle, \quad |b_6\rangle = |1'_A 0'_B 1'_C\rangle, \quad |b_7\rangle = |1'_A 1'_B 0'_C\rangle, \quad |b_8\rangle = |1'_A 1'_B 1'_C\rangle. \tag{7.25}$$

The vectors $|b_1\rangle$ to $|b_4\rangle$ span a subspace $\mathcal{G}$ in $\mathcal{H}^{\otimes 3}$, and the probability that $|\Psi\rangle$ belongs to $\mathcal{G}$ is

$$\mathsf{p}_\mathcal{G} = \beta^6 + 3\beta^4\gamma^2 \simeq 0.9419.$$

Therefore, it is much more likely (roughly 20 times more likely) that the message belongs to $\mathcal{G}$ rather than to its orthogonal complement $\mathcal{G}_\perp$. Since $\mathcal{G}$ is a four-dimensional space, two qubits should be enough to send the message with a very good fidelity.

In practice, Alice can use a unitary transformation U which rotates the four basis vectors $|b_1\rangle \cdots |b_4\rangle$ of $\mathcal{G}$ into basis states of the form $|\varphi_A \varphi_B 0_C\rangle$ and the four basis vectors $|b_5\rangle \cdots |b_8\rangle$ of $\mathcal{G}_\perp$ into basis states $|\overline{\varphi}_A \overline{\varphi}_B 1_C\rangle$. She measures qubit C, and if the outcome is 0 she sends the remaining two qubits to Bob. The compressed message Ψ_{comp} is given by

$$|\Psi_{\mathrm{comp}} \otimes 0_C\rangle = U|\Psi\rangle, \qquad |\Psi\rangle \in \mathcal{G}.$$

Bob receives the two-qubit message, takes its tensor product with $|0_C\rangle$ and he reads it by applying U^{-1}:

$$|\Psi'\rangle = U^{-1}|\Psi_{\mathrm{comp}} \otimes 0_C\rangle \equiv |\Psi\rangle.$$

If Alice's measurement of qubit C gives $|1\rangle$, then the best she can do is send Bob the state that he will decompress to the most likely state $|0'_A 0'_B 0'_C\rangle$, that is, she sends the state $|\Psi_{\mathrm{comp}}\rangle$ such that

$$|\Psi'\rangle = U^{-1}|\Psi_{\mathrm{comp}} \otimes 1_C\rangle = |0'_A 0'_B 0'_C\rangle.$$

The outcome of the procedure is that Bob obtains the state matrix

$$\sigma' = \mathcal{P}|\Psi\rangle\langle\Psi|\mathcal{P} + |b_1\rangle\langle\Psi|(I - \mathcal{P})|\Psi\rangle\langle b_1|, \tag{7.26}$$

where $\mathcal{P}$ is the projector onto $\mathcal{G}$.

Let us now proceed to the general case. The key to Schumacher's theorem is that it is sufficient to encode typical subspaces of $\mathcal{H}^{\otimes n}$ if we wish to send n qubits drawn from the ensemble $\{\mathsf{p}_\alpha, |\alpha\rangle\}$ which defines the state matrix (7.13). Since the letters are drawn independently, the state matrix of the n qubits is

$$\rho^{\otimes n} = \rho_1 \otimes \cdots \otimes \rho_n = \sigma. \tag{7.27}$$

Now, each ρ_i can be written in diagonal form (7.14). The eigenvalues of ρ_i are $\lambda_1 = \mathsf{p}$ and $\lambda_2 = 1 - \mathsf{p}$. An eigenvalue of $\rho^{\otimes n}$ will be of the form

$$\lambda_1^q \lambda_2^{n-q} = \mathsf{p}^q (1-\mathsf{p})^{n-q}$$

and it will appear $\binom{n}{q}$ times. From our preceding discussion of Shannon's theorem, we see that almost all the eigenvalues of $\rho^{\otimes n}$ lie in a domain defined by the following range of q:

$$n\mathsf{p} - \mathcal{O}(\sqrt{n}) \lesssim q \lesssim n\mathsf{p} + \mathcal{O}(\sqrt{n}).$$

For these values of q the eigenvalues of $\rho^{\otimes n}$ will be

$$\lambda \simeq 2^{-nH_{\mathrm{Sh}}(\mathsf{p})} = 2^{-nH_{\mathrm{vN}}(\rho)},$$

with $H_{\mathrm{vN}}(\rho) = -\sum_i \mathsf{p}_i \ln \mathsf{p}_i$. In other words, the typical eigenvalue of $\rho^{\otimes n}$ is $\lambda = 2^{-nH_{\mathrm{vN}}(\rho)}$. Let $\mathcal{G}$ be the subspace of $\mathcal{H}^{\otimes n}$ spanned by the eigenvectors corresponding to these eigenvalues, and let $\mathcal{P}$ be the projector onto this subspace. Then for any $\varepsilon > 0$ we can choose n large enough that

$$\mathrm{Tr}\left(\rho^{\otimes n}\mathcal{P}\right) \geq 1 - \varepsilon. \tag{7.28}$$

Suppose that Alice wants to send an n-letter message drawn from the ensemble $\{\alpha\}$:

$$|\Psi\rangle = |\alpha_1 \cdots \alpha_n\rangle.$$

As above, she uses a unitary transformation U such that its action on a typical message belonging to $\mathcal{G}$ is

$$U|\Psi_{\mathrm{typ}}\rangle = |\Psi_{\mathrm{comp}} \otimes 0 \otimes \cdots \otimes 0\rangle.$$

Then she sends Bob the nH_{vN} qubits corresponding to the space $\mathcal{G}$ and Bob decodes it using U^{-1}. The state received by Bob is

$$\sigma' = \mathcal{P}|\Psi\rangle\langle\Psi|\mathcal{P} + \overline{\sigma}', \tag{7.29}$$

where $\overline{\sigma}'$ is what Alice sends if $|\Psi\rangle \notin \mathcal{G}$. The fidelity of σ' obeys the inequality

$$\mathcal{F}(|\Psi\rangle\langle\Psi|, \sigma') \geq |\langle\Psi|\mathcal{P}|\Psi\rangle|^2 \tag{7.30}$$

because $\overline{\sigma}'$ is a positive operator. The fidelity depends on the message $|\alpha_1 \cdots \alpha_n\rangle$, and to obtain the final result we must sum over the p_α:

$$\mathcal{F}(\sigma, \sigma') \geq \sum_\alpha \mathsf{p}_\alpha |\langle\Psi|\mathcal{P}|\Psi\rangle|^2$$

$$\geq \sum_\alpha \mathsf{p}_\alpha (2\langle\Psi|\mathcal{P}|\Psi\rangle - 1) \tag{7.31}$$

$$= 2\mathrm{Tr}\left(\rho^{\otimes n}\mathcal{P}\right) - 1 \geq 1 - \varepsilon,$$

where we have used $x^2 \geq 2x - 1$ and (7.28). In order to obtain a fidelity arbitrarily close to one, it is enough to send nH_{vN} qubits when $n \to \infty$.

7.4 Quantum error correction

Noise is omnipresent in all classical data processing, communication, and storage and it introduces errors. For example, noise can flip the initial value 0 of a bit to the value 1, something we wish to avoid so that all our operations do not fall apart. The task of error correction is to detect the erroneous bits and correct them. Modern (classical) error-correcting codes are extremely sophisticated in their details, but they are all based on redundancy. A very simple example is the following. Instead of encoding information in a single bit, we encode it in three bits:

$$0 \rightarrow 000, \qquad 1 \rightarrow 111.$$

Suppose that the effect of noise is to flip a bit from 0 to 1 or vice versa with probability p. Then if we start with the bits in the state 000, for example, a single-bit flip will occur with probability $3p(1-p)^2$ and a two-bit flip with probability $3p^2(1-p)$. If we read

$$000, \qquad 100, \qquad 010, \qquad \text{or} \qquad 001$$

we can decide with probability $(1-p)^2(1+2p)$ that the original bit had the value 0. Therefore, a majority rule gives the correct result with probability $(1-p)^2(1+2p)$. If $p = 10^{-2}$, the probability of error is $p^2(3-2p) \simeq 3 \times 10^{-4}$, while it would be 10^{-2} without corrrection.

Classical error correction cannot be transposed directly to qubits for four reasons.

1. The no-cloning theorem forbids duplicating qubits in an unknown state.
2. There is no classical analog to the superposition of qubits, so that a phase flip such as (4.25) has no classical equivalent.
3. Errors may be continuous. For example, in the general qubit state

$$|\varphi\rangle = \cos\frac{\theta}{2}|0\rangle + e^{i\phi}\sin\frac{\theta}{2}|1\rangle$$

   noise could lead to continuous variations of the angles θ and ϕ.
4. Finally, measurement destroys quantum information: one should not affect the information encoded in a qubit by a projective measurement.

In spite of these difficulties, it has been possible to devise quantum error-correcting codes. These codes are rather involved, and we shall limit ourselves to a simple but illustrative example for the case of the phase flip σ_z introduced in (4.25):

$$\lambda|0\rangle + \mu|1\rangle \rightarrow \sigma_z(\lambda|0\rangle + \mu|1\rangle). \tag{7.32}$$

This phase flip is a typical quantum error, because a classical bit cannot be in a linear superposition. It will be convenient to rephrase (7.32) by going to the $|\pm\rangle$ basis:

$$|\pm\rangle = \frac{1}{\sqrt{2}}\left(|0\rangle \pm |1\rangle\right). \qquad (7.33)$$

We recall that the $|\pm\rangle$ vectors are obtained from $|0\rangle$ and $|1\rangle$ by application of the H gate:

$$H|0\rangle = |+\rangle, \qquad H|1\rangle = |-\rangle.$$

Then a phase flip in the $\{|0\rangle, |1\rangle\}$ basis corresponds to a bit flip in the $|\pm\rangle$ basis:

$$H\left(\lambda|0\rangle + \mu|1\rangle\right) = \lambda|+\rangle + \mu|-\rangle,$$
$$\sigma_z\, H\left(\lambda|0\rangle + \mu|1\rangle\right) = \lambda|-\rangle + \mu|+\rangle. \qquad (7.34)$$

The redundancy is introduced by using, in addition to the original qubit A, two auxiliary qubits B and C in the state $|0\rangle$:

$$\left(\lambda|0\rangle + \mu|1\rangle\right) \otimes |00\rangle = \lambda|000\rangle + \mu|100\rangle,$$

to which we apply two cNOT gates controlled by qubit A (left-hand side of the circuit drawn in Fig. 7.3)

$$\text{cNOT}_B\,\text{cNOT}_C\,(\lambda|000\rangle + \mu|100\rangle) = \lambda|000\rangle + \mu|111\rangle, \qquad (7.35)$$

followed by three Hadamard gates

$$H^{\otimes 3}\,(\lambda|000\rangle + \mu|111\rangle) = \lambda|+++\rangle + \mu|---\rangle = |\Psi_0\rangle. \qquad (7.36)$$

In the absence of any phase flip, the final state of the three qubits is $|\Psi_0\rangle$ (7.36). If the phase of one of the three qubits is flipped, we get

$$|\Psi_A\rangle = \lambda|-++\rangle + \mu|+--\rangle \qquad \text{qubit } A \text{ flipped}, \qquad (7.37)$$
$$|\Psi_B\rangle = \lambda|+-+\rangle + \mu|-+-\rangle \qquad \text{qubit } B \text{ flipped}, \qquad (7.38)$$
$$|\Psi_C\rangle = \lambda|++-\rangle + \mu|--+\rangle \qquad \text{qubit } C \text{ flipped}. \qquad (7.39)$$

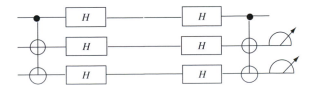

Figure 7.3 Circuit for error-correcting code.

Let us introduce the two operators $X_A X_B$ and $X_A X_C$, where $X \equiv \sigma_x$. The vectors $|\Psi_0\rangle \cdots |\Psi_C\rangle$ are eigenvectors of these two operators with eigenvalues $+1$ or -1:

$$X_A X_B |\Psi_0\rangle = +|\Psi_0\rangle, \quad X_A X_C |\Psi_0\rangle = +|\Psi_0\rangle, \quad X_A X_B |\Psi_A\rangle = -|\Psi_A\rangle,$$

$$X_A X_C |\Psi_A\rangle = -|\Psi_A\rangle, \quad X_A X_B |\Psi_B\rangle = -|\Psi_B\rangle, \quad X_A X_C |\Psi_B\rangle = +|\Psi_B\rangle, \quad (7.40)$$

$$X_A X_B |\Psi_C\rangle = +|\Psi_C\rangle, \quad X_A X_C |\Psi_C\rangle = -|\Psi_C\rangle.$$

The measurement of $X_A X_B$ and $X_A X_C$ allows us to determine the type of error which has occurred:

$$
\begin{aligned}
X_A X_B &= +1, & X_A X_C &= +1 & &\text{no error,} \\
X_A X_B &= -1, & X_A X_C &= -1 & &\text{bit } A \text{ flipped,} \\
X_A X_B &= -1, & X_A X_C &= +1 & &\text{bit } B \text{ flipped,} \\
X_A X_B &= +1, & X_A X_C &= -1 & &\text{bit } C \text{ flipped.}
\end{aligned}
\qquad (7.41)
$$

However, the qubits should not be measured individually, as this would lead to the destruction of information on qubit A. The measurement is performed according to the right-hand side of the circuit in Fig. 7.3. For example,

$$
\begin{aligned}
\mathrm{cNOT}_B \, \mathrm{cNOT}_C (H_A \otimes H_B \otimes H_C) |\Psi_B\rangle &= \mathrm{cNOT}_B \, \mathrm{cNOT}_C \, (\lambda|010\rangle + \mu|110\rangle) \\
&= \lambda|010\rangle + \mu|110\rangle \\
&= (\lambda|0\rangle + \mu|1\rangle) \otimes |10\rangle.
\end{aligned}
\qquad (7.42)
$$

If the qubits B and C are found in the states $|1\rangle$ and $|0\rangle$, respectively, this implies that qubit B was flipped. The reader will easily check (Exercise 7.5.5) that the final states of the qubits B and C are

$$|\Psi_0\rangle \to |00\rangle, \quad |\Psi_A\rangle \to |11\rangle, \quad |\Psi_B\rangle \to |10\rangle, \quad |\Psi_C\rangle \to |01\rangle.$$

If the measured values of qubits B and C give the state $|11\rangle$, then we apply X_A to qubit A. The correct quantum state is therefore recovered without ever measuring this qubit: the measurement does not give any information on the values of λ and μ.

There are other types of error in addition to phase flip. In order to deal with all the errors it is necessary to use at least four auxiliary qubits, but this so-called five-qubit correcting code is extremely cumbersome. At present the most favored code is the seven-qubit correcting code devised by Sheane, although the first code devised by Shor, which is a nine-qubit code, also has interesting properties.

7.5 Exercises

7.5.1 Superdense coding

Alice and Bob share a pair of entangled qubits A and B in state $|\Psi\rangle$ (see Fig. 7.4)

$$|\Psi\rangle = \frac{1}{\sqrt{2}}\left(|0_A \otimes 0_B\rangle + |1_A \otimes 1_B\rangle\right).$$

Alice wishes to send Bob two *classical* bits of information i and j, $i, j = 0, 1$, while using a single qubit. She transforms the state of her qubit by applying on it the operator A_{ij} acting on qubit A

$$A_{ij} = (\sigma_{xA})^i(\sigma_{zA})^j$$

where i and j are exponents. She then sends her qubit to Bob, who gets the pair in the state $A_{ij}|\Psi\rangle$.

1. Give the explicit expression of $A_{00}|\Psi\rangle$, $A_{01}|\Psi\rangle$, $A_{10}|\Psi\rangle$, $A_{11}|\Psi\rangle$ in terms of the states $|0_A \otimes 0_B\rangle$, $|0_A \otimes 1_B\rangle$, $|1_A \otimes 0_B\rangle$, $|1_A \otimes 1_B\rangle$.
2. Bob uses the logic circuit of Fig. 7.4 with a cNOT gate and a Hadamard gate H. Examining the four possibilities for $A_{ij}|\Psi\rangle$, show that the cNOT gate transforms $A_{ij}|\Psi\rangle$ into a tensor product and that measurement of qubit B gives the value of i. Show finally that measurement of qubit A gives the value of j. Thus Alice transmits two bits of information while sending only one qubit.

7.5.2 Shannon entropy versus von Neumann entropy

Let us consider a two-dimensional space and define the state $|\theta\rangle$ as

$$|\theta\rangle = \cos\frac{\theta}{2}|0\rangle + \sin\frac{\theta}{2}|1\rangle.$$

Let a state matrix ρ be given by

$$\rho = \mathsf{p}|0\rangle\langle 0| + (1-\mathsf{p})|\theta\rangle\langle\theta|.$$

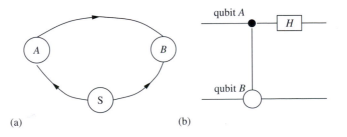

(a) (b)

Figure 7.4 (a) General depiction; S, source of entangled particles. (b) Gates applied by Bob.

Compute the Shannon and von Neumann entropies. Show that

$$H_{\mathrm{Sh}} \geq H_{\mathrm{vN}}.$$

7.5.3 Information gain of Eve

Let us compute the information gain $I(\alpha : \varepsilon)$ (see (7.11)) of Eve, where α stands for Alice and ε for Eve, in two different situations.

1. Let i characterize the bit sent by Alice in the $\{|x\rangle, |y\rangle\}$ basis or in the $\{|\pi/4\rangle,$ $|-\pi/4\rangle\}$ basis. Thus i can take four different values with equal probabilities $\mathsf{p}(i) = 1/4$. Let Eve use the $\{|x\rangle, |y\rangle\}$ basis in which she measures a result r, where r takes two different values. Establish a table of the conditional probabilities $\mathsf{p}(r|i)$ and deduce from it $\mathsf{p}(i|r)$. Show that Eve's information gain is 1/2.
2. Now Eve uses a symmetric $\{|\pi/8\rangle, |-\pi/8\rangle\}$ basis (see Exercise 2.6.4). Show that in this case the information gain is only $I(\alpha : \varepsilon) \simeq 0.4$. The information gain is smaller when Eve uses the symmetric basis.

7.5.4 Symmetry of the fidelity

Show that

$$\mathcal{F}(\rho, |\Psi\rangle\langle\Psi|) = \mathcal{F}(|\Psi\rangle\langle\Psi|, \rho) \ .$$

Hint: to evaluate the first expression for $\mathcal{F}$, use a basis (7.14) where ρ is diagonal, and observe that a product of matrices is of rank one if one of the matrices in the product is of rank one. What is then the nonzero eigenvalue of the product

$$\rho^{1/2}|\Psi\rangle\langle\Psi|\rho^{1/2}?$$

It may be instructive to examine first the case of two-dimensional matrices.

7.5.5 Quantum error correcting code

Work out the details of the calculations leading to (7.41) and the action of the transformation on the right-hand side of the circuit in Fig. 7.3 in the four different cases.

7.6 Further reading

Zeilinger (2000) gives an elementary account of teleportation. Recent experiments demonstrating teleportation using atoms are described by Barret *et al.* (2004) and

Riebe *et al.* (2004). Shannon entropy, von Neumann entropy and Schumacher theorem are explained by Preskill (1999), Chapter 5, Nielsen and Chuang (2000), Chapters 11 and 12 or Stolze and Suter (2004), Chapter 13. For quantum error correction, see Nielsen and Chuang (2000), Chapter 10 or Stolze and Suter (2004), Chapter 7.

References

Arndt, M., Hornberger, K. and Zeilinger, A. (2005), Probing the limits of the quantum world, *Physics World* **18**, 2005, 35–40.

Aspect, A. (1999), Bell's inequalities: more ideal than ever, *Nature* **398**, 189–190.

Aspect, A. and Grangier, Ph. (2004), Des intuitions d'Einstein aux bits quantiques, *Pour la Science* **326**, December, 120–125.

Badurek, G., Rauch, H. and Tuppinger, D. (1985), Neutron interferometric double-resonance experiment, *Physical Review A* **34**, 2600–2608.

Barrett, M. *et al.* (2004), Deterministic quantum teleportation of atomic qubits, *Nature* **429**, 737–739.

Bennett, C. H. (1987), Demons, engines, and the second law, *Scientific American*, November, 88–96.

Bennett, C., Brassard, G. and Ekert, A. (1992), Quantum cryptography, *Scientific American*, 1992, 26–33.

Blatt, R. (2004), Quantum information processing in ion traps, in *Les Houches Summer School 2003, Quantum Entanglement and Information Processing*, ed. D. Estève, J.-M. Raimond, and M. Brune, Amsterdam: Elsevier, pp. 223–260.

Bouwmeester, D., Ekert, A. and Zeilinger, A. (2000), *The Physics of Quantum Information*, Berlin: Springer.

Burkard, G., Engel, H. A. and Loss, D. (2002), Spintronics, quantum computing and quantum communication in quantum dots, in *Fundamentals of Quantum Information*, ed. D. Heiss, Lecture Notes in Physics **587**, Berlin: Springer, pp. 241–264.

Cirac, J. and Zoller, P. (2004), New frontiers in quantum information with atoms and ions, *Physics Today* **57**, March, 38–44.

Cohen-Tannoudji, C., Diu, B. and Laloë, F. (1977), *Quantum Mechanics*, New York: Wiley.

Devoret, M. and Martinis, J. (2004), Superconducting qubits, in *Les Houches Summer School 2003, Quantum Entanglement and Information Processing*, ed. D. Estève, J.-M. Raimond, and M. Brune, Amsterdam: Elsevier, pp. 443–485.

Dürr, S., Nonn, T. and Rempe, G. (1998), Origin of quantum-mechanical complementarity probed by a "which way" experiment in an atom interferometer, *Nature* **395**, 33–37.

Englert, B., Scully, M. and Walther, H. (1991), Quantum optical tests of complementarity, *Nature* **351**, 111–116.

Ekert, A. and Josza, R. (1996), Shor's factoring algorithm, *Reviews of Modern Physics* **68**, 733–753.

Einstein, A., Podolsky, B. and Rosen, N. (1935), Can quantum-mechanical description of physical reality be considered complete?, *Physical Review* **47**, 777–780.

Hanson, R. *et al.* (2005) Single shot readout of electron spin states in a quantum dot using spin dependent tunnel rates, *Physical Review Letters* **94**, 1968021–4.

Hey, T. and Walters, P. (2003), *The New Quantum Universe*, Cambridge: Cambridge University Press.

Gisin, N., Ribordy, G., Tittel, W. and Zbinden, H. (2002), Quantum cryptography, *Reviews of Modern Physics* **74**, 145–195.

Johnson, G. (2003), *A Shortcut Through Time: the Path to the Quantum Computer*, New York: Knopf.

Jones, J. (2004), Nuclear magnetic resonance computation, in *Les Houches Summer School 2003, Quantum Entanglement and Information Processing*, ed. D. Estève, J.-M. Raimond, and M. Brune, Amsterdam: Elsevier, pp. 357–400.

Landauer, R. (1991a), The physical nature of information, *Physics Letters A* **217**, 188–193.

Landauer, R. (1991b), Information is physical, *Physics Today*, May, 23–29.

Le Bellac, M. (2006), *Quantum Physics*, Cambridge: Cambridge University Press.

Leggett, A. (2002), Qubits, cbits, quantum measurement and environment, in *Fundamentals of Quantum Information*, ed. D. Heiss, Lecture Notes in Physics **587**, Berlin: Springer, pp. 3–46.

Leibfried, D., Blatt, R., Monroe, C. and Wineland, D. (2003), Quantum dynamics of trapped ions, *Reviews of Modern Physics* **75**, 281–324.

Levitt, M. H. (2001), *Spin Dynamics, Basics of Nuclear Magnetic Resonance*, New York: Wiley.

Lévy-Leblond, J.-M. and Balibar, F. (1990), *Quantics: Rudiments of Quantum Physics*, New York: North-Holland.

Mermin, N. (1993), Hidden variables and the two theorems of John Bell, *Reviews of Modern Physics* **65**, 803.

Mermin, N. (2003), Lecture notes in quantum computation, http://www.ccmr.cornell.edu/~mermin/qccomp/CS483.html.

Mooij, H. (2004), Superconducting quantum bits, *Physics World* **17**, December, 29–33.

Nielsen, M. and Chuang, I. (2000), *Quantum Computation and Quantum Information*, Cambridge: Cambridge University Press.

Omnès, R. (1994), *The Interpretation of Quantum Mechanics*, Princeton, NJ: Princeton University Press.

Paz, J. and Zurek, W. (2002), Environment induced decoherence and the transition from quantum to classical, in *Fundamentals of Quantum Information*, ed. D. Heiss, Lecture Notes in Physics **587**, Berlin: Springer, pp. 77–148.

Peres, A. (1993), *Quantum Theory, Concepts and Methods*, Boston, MA: Kluwer.

Preskill, J. (1999), Quantum computing, http://www.theory.caltech.edu/~preskill/.

Riebe, M. *et al.* (2004), Deterministic quantum teleportation with atoms, *Nature* **429**, 734–737.

Singh, S. (2000), *The Code Book: The Science of Secrecy from Ancient Egypt to Quantum Cryptography*, New York: Anchor Books.

Stolze, J. and Suter, D. (2004), *Quantum Computing*, New York: Wiley.

Vandersypen, L. and Chuang, I. (2005), NMR techniques for quantum control and computation, *Reviews of Modern Physics* **76**, 1037–1069.

Vandersypen, L., Steffen, M., Breyta, G., Yannoni, C., Sherwood, M. and Chuang, I. (2001), Experimental realization of quantum Shor's factoring algorithm using nuclear magnetic resonance, *Nature* **414**, 883–887.

di Vincenzo, D. (2000), The physical implementation of quantum computation, *Fortsschritte der Physik* **48**, 771–783.

You, J. and Nori, F. (2005), Superconducting circuits and quantum information, *Physics Today*, November, 42–47.

Zeilinger, A. (2000), Quantum teleportation, *Scientific American*, April, 32–41.

Zurek, W. (1991), Decoherence and the transition from quantum to classical, *Physics Today*, October, 36–44.

Index